MATHEMATICS IN MICROBIOLOGY

MATHEMATICS IN MICROBIOLOGY

Edited by

M. BAZIN
*Department of Microbiology,
Queen Elizabeth College,
University of London, UK*

ACADEMIC PRESS
A Subsidiary of Harcourt Brace Jovanovich, Publishers
London   New York
Paris   San Diego   San Francisco   Saō Paulo
Sydney   Tokyo   Toronto

ACADEMIC PRESS INC. (LONDON) LTD.
24/28 Oval Road
London NW1 7DX

United States Edition published by
ACADEMIC PRESS INC.
111 Fifth Avenue
New York, New York 10003

Copyright © 1983 by ACADEMIC PRESS INC. (LONDON) LTD.

All rights Reserved
No part of this book may be reproduced in any form by photostat, microfilm,
or any other means, without written permission from the publishers

British Library Cataloguing in Publication Data
Mathematics in microbiology
 1. Microbiology—Mathematics
 I. Bazin, M.
 576   QH434
 ISBN 0-12-083480-4

LCCCN 82-074352

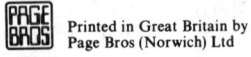
Printed in Great Britain by
Page Bros (Norwich) Ltd

Contributors

W. G. Characklis Department of Civil Engineering and Engineering Mechanics, Montana State University, Bozeman 59717, Montana, U.S.A.

J. M. Crowther Department of Applied Physics, John Anderson Building, University of Strathclyde, 107 Rottenrow, Glasgow G4 0NG, U.K.

A. Cunningham Department of Applied Physics, John Anderson Building, University of Strathclyde, 107 Rottenrow, Glasgow G4 0NG, U.K.

J. F. Dalrymple Department of Management Science and Technology Studies, University of Stirling, Stirling FK9 4LA, U.K.

A. Ebringer Immunology Unit, Departments of Biochemistry and Microbiology, Queen Elizabeth College, London W8 7AH, U.K.

J. D. Harvey Department of Physics, University of Auckland, Private Bag Auckland, New Zealand.

M. A. Holder-Franklin Environmental Microbiology Research Laboratory, University of Windsor, 401 Sunset Avenue, Windsor, Ontario N9B 3P4, Canada.

J. A. Howell Department of Chemical Engineering, University College of Swansea, Singleton Park, Swansea, U.K.

R. M. Nisbet Department of Applied Physics, John Anderson Building, University of Strathclyde, 107 Rottenrow, Glasgow G4 0NG, U.K.

P. T. Saunders Department of Mathematics, Queen Elizabeth College, London W8 4AH, U.K.

L. J. Wuest Environmental Research Laboratory, University of Windsor, 401 Sunset Avenue, Windsor, Ontario N9B 3P4, Canada.

To Marian

Preface

In this volume I have attempted to collect together a series of articles which illustrate the way some branches of mathematics have been used to investigate specific problems in microbiology and the related field of immunology. In all cases technical details have been kept to a minimum but sufficient information has been retained in the mathematical aspects to supply the reader with an understanding of the methods involved, and the biological portions of each chapter are treated in some detail.

The text is directed towards biological scientists who have some familiarity with differential equations at a level obtained in most introductory calculus courses. Hopefully, some physical scientists might also be attracted to the text in which case it is hoped that the description of the biological material will be sufficient for their purposes.

Each chapter is preceded by a brief introduction which is designed to set the scene for what follows. The book begins by considering mathematical aspects of the basic unit of biology, the cell, and then proceeds with an introduction to modelling dynamic systems. The analysis of population models is continued in Chapter 3 and then a new theme is introduced by P. T. Saunders in a chapter entitled "Catastrophe Theory". The analysis of large data sets by factor analysis is dealt with next and there follows a chapter on process analysis in which the rigorous approach of the chemical engineer is exemplified. The penultimate chapter deals with time-domain analysis and, finally, information theory is used to refute an established theory of antibody diversity.

I would like to thank all the contributors to this volume for their perseverance and patience during its preparation and also my daughters, Margaret and Victoria, whose support provides a basis for all my endeavours.

London   M. BAZIN
1983

Contents

List of Contributors v
Preface vii

Chapter 1
Mathematics of Microbial Age and Size Distributions
J. D. HARVEY

1. Introduction 1
2. Size and Age Distributions 2
   2.1. Notation and Ideas of Probablity Theory 3
   2.2. Analytical Forms for Probability Distributions . . . 7
3. Steady-State Growth 8
4. Relationships between Distributions 10
   4.1. Relationships between Generation-Time Distributions . . 11
   4.2. Relationships between the Size Distributions . . . 15
   4.3. Relationships between Size and Generation-Time Distributions . 17
   4.4. The Age Distribution 19
5. Correlations between Variables Associated with the Cell Cycle . . 20
   5.1. The Development of Synchronous Cultures 22
6. Experimental Results for Size and Age Distributions . . . 26
7. Specific Models for Cell Growth and Division 28
8. Summary 34
References 35

Chapter 2
Mathematical Models in Microbiology: Mathematical Tool-Kit
J. A. HOWELL

1. Introduction 37
2. Kinetic Models 38
   2.1. Exponential Growth 38
   2.2. Logistic Growth 40
   2.3. Continuous-Flow Stirred Fermenter 40
   2.4. Substrate Limitation 42
   2.5. Multiple Substrate Kinetics 43
   2.6. Structured Models 44
3. Mass Balances 46
   3.1. Stoichiometry 46
   3.2. Transport Terms 47
   3.3. Diffusion Barriers 48
   3.4. Solutions 51
   3.5. Estimating Model Parameters 52
   3.6. System Reduction 55

4. Linearization of Non-Linear Differential Equations . . . 56
    4.1. Steady States 56
    4.2. Taylor Series Expansion 56
    4.3. Linearization about a Steady State 58
    4.4. Mean Value Linearization 59
5. Mathematical Representation 60
    5.1. Dimensionless Equations 60
    5.2. Matrix/Vector Notation 61
6. Systems of Linear Differential Equations 63
    6.1. Solution 63
    6.2. Significance of the Eigenvalues 64
    6.3. Stability 67
    6.4. Examples 70
References 75

Chapter 3
Transients and Oscillations in Continuous Culture
A. CUNNINGHAM AND R. M. NISBET

1. Introduction 77
2. Mathematical Methods 78
    2.1. Differential-Equation Models 78
    2.2. The Effect of Time Delays 82
3. The Transient Behaviour of Simple Models of Microbial Growth and Predation 84
    3.1. Models of the Growth of a Single Species . . 84
    3.2. Models of Microbial Predation 91
    3.3. The Use of Operating Diagrams to Illustrate System Behaviour . 97
4. Transients and Oscillations Observed Experimentally . . 99
    4.1. Single Species 99
    4.2. Prey and Predator 100
5. Conclusions 102
References 102

Chapter 4.
Catastrophe Theory
P. T. SAUNDERS

1. Introduction 105
2. The Zeeman Catastrophe Machine 106
3. Catastrophe Theory 111
4. Structural Stability 113
5. An Example of the Fold Catastrophe 115
6. Two Applications of Catastrophe Theory 119
    6.1. Determination of Critical Variables . . . 121
    6.2. Culmination 130
7. A Note on Qualitative Results 135
8. Conclusion 136
References 138

Chapter 5
Factor Analysis as an Analytical Method in Microbiology
MAXINE A. HOLDER-FRANKLIN AND L. J. WUEST

1. Introduction 139
2. Factor Analysis and the Selection of the Database 140
   2.1. Factor Analysis 140
   2.2. Numerical Taxonomy; the Selection of the Database . . 142
3. Population Shifts in Heterotrophic Bacteria 142
   3.1. Materials and Methods 145
   3.2. Factor Analytic Results 152
4. Interpretation of Factors 154
   4.1. Microbiological Factors 154
   4.2. Physico-Chemical Factors (Naquadat) 156
   4.3. Interpretation of Microbial Factor Scores . . . 157
   4.4. Physico-Chemical Naquadat (NQ) Factor Scores . . 158
   4.5. Correlation of Physico-Chemical with Biological Factors . 159
   4.6. Diurnal and Seasonal Changes in the Predominant Bacterial Species 164
   4.7. Correlations of Biological Factors with Environmental Factors . 165
5. Summary 168
References 168

Chapter 6
Process Analysis in Microbial Systems: Biofilms as a Case Study
W. G. CHARACKLIS

1. Introduction 171
2. Process Analysis 174
   2.1. Modelling 174
   2.2. Fundamentals Regarding Rate 178
   2.3. Reactors 181
   2.4. Reaction Kinetics 187
3. Modelling of Microbial Processes 194
   3.1. Modelling Variables 195
   3.2. Physical Principles 195
   3.3. Biological Principles 196
   3.4. Simplifying Assumptions 197
4. Biofilm Processes 200
   4.1. Properties and Composition of Biofilms 201
   4.2. Rate Processes Contributing to Biofilm Development . . 208
   4.3. Effects of Biofilms on Fluid Frictional Resistance . . 223
   4.4. Effects of Biofilms on Heat-Transfer Resistance . . 229
5. Summary 232
Acknowledgements 232
References 232

Chapter 7
Theory and Practice of Time-Domain Techniques
J. F. DALRYMPLE AND J. M. CROWTHER

1. Introduction	235
2. Time-Domain Techniques	239
2.1. Introduction	239
2.2. The Box and Jenkins Techniques	240
3. Application of Time-Domain Techniques	258
3.1. General Considerations	258
3.2. A Worked Example	261
4. Conclusions	284
References	284

Chapter 8
The Random Somatic Mutation Theory of Antibody Diversity and Information Theory
A. EBRINGER

1. Introduction	287
2. Selective Theories and Antibody Diversity	289
3. Immune Response as Signal Transmission	290
4. Information Theory	291
5. The Random Somatic Mutation Theory of Antibody Diversity	294
6. Information Theory and Cross-Reactivity	297
7. Linguistics and Cross-Reactivity	301
8. Summary and Conclusion	302
References	303
Index	305

Chapter 1

Mathematics of Microbial Age and Size Distributions

J. D. Harvey

Department of Physics, University of Auckland, Private Bag, Auckland, New Zealand

The properties of a microbial population are determined by the cells that comprise it. Therefore, if the behaviour of the individual cells is known it should be possible to predict the behaviour of the population. In this first chapter the author uses the converse of this argument to investigate size and age relationships in cells using aspects of probability theory applied to population data—*Editorial note.*

1. Introduction

A substantial body of literature has been developed in the past 25 years, concerning the measurement and analysis of the sizes of microbial cells and of their generation times. There are two main origins of the interest in these variables associated with the cell cycle. In the first place microbial populations are often of such a size that the probability of a particular cell mass or interdivision time occurring can be measured readily from a determination of the proportion of organisms having such an attribute. The size and age distributions thus constitute well-defined parameters of the culture which are amenable both to direct experimental measurement and to rigorous mathematical analysis. The second reason for the interest in these distributions lies in their ability to test models which can be proposed for the mechanism of growth and cellular division. Model building in biology is of little value unless the models can be tested in well-designed experiments. The precise identification of the parameters of the size and generation-time distributions has led to the rejection of a number of models governing the size and age of cells at division.

As experimental methods have improved more sophisticated models and methods of analysis of the distributions have been developed. The appearance of a simple pattern in the distribution of generation times, for example, is analogous to the appearance of macroscopic laws in the physical sciences which can be derived from more fundamental microscopic laws. As an example the macroscopic laws of diffusion can be derived from the microscopic laws governing the collisions of molecules. The value of this derivation lies in the deeper understanding of the process which is thereby attained, and it can be hoped that an understanding of the properties of microbial age and size distributions will ultimately be obtained in terms of the chemistry and physics governing the growth process. Such an analysis is not at this stage possible, and the very complexity of the phenomena which are being described may well make such an understanding unattainable in the foreseeable future. The value of the distributions themselves is not diminished by this difficulty, however, and it is important that they be as well defined as is mathematically and experimentally feasible.

Much of the mathematics necessary to describe the variation in cell sizes and ages is related to that used in other disciplines. The level of mathematical sophistication at which the subject is best approached is thus determined by the experience of the investigator, but many of the more significant results can be derived without resorting to the use of sophisticated mathematical tools, and this is the approach used in this chapter. A sound theoretical (i.e. mathematical) understanding of the statistics of the variables associated with a microbial culture is essential for an accurate prediction of the growth of cellular components and of cell numbers in a culture. This understanding involves an appreciation of the mathematical consequences of the steady state of growth, of the correlations between variables associated with the cell cycle and of the interrelations between the various distributions of random variables which occur in the population. These topics are dealt with in the ensuing sections after first establishing the notation used in the mathematical formulation.

2. Size and Age Distributions

It has long been recognized that both the size (as determined either by the volume or by a linear dimension) and the interdivision time of cells selected from a culture exhibit substantial variability. The distribution of cell sizes can be obtained experimentally by measuring the sizes of a sufficiently large number of cells selected without bias from the initial population, and constructing a histogram of the number of cells in each measured size class. It is reasonable to assume that the probability distribution function for the size of a single cell drawn from the culture, is a smooth continuous function, although the measured histogram may be far from smooth due to the experimental difficulties involved.

In order to improve the measured distribution, it is necessary to divide the size classes into smaller and smaller intervals and to perform sufficient measurements to obtain a good estimate of the probability of cells in a given size class occurring. Equivalently, it becomes necessary to count larger and larger numbers of cells. Methods involving human intervention limit the total number of cells which can be counted to numbers of the order of thousands. Measurements of cell sizes and numbers can, however, be performed by automatic cell counters, and the rate of measurement using such instruments is so high that the measured size distribution is no longer limited by the number of cells counted, but by the inherent precision of measurement of the instrument.

When calculating the rate of synthesis of some biochemical component of a cell, or the distribution of cellular components between individual cells in a culture, one is led naturally to a consideration of the distribution of ages of cells in a culture. In considering the mechanism for the control of cellular division one is led similarly to the discussion of the distribution of interdivision times within the culture. A quantitative description of these distributions can only be obtained mathematically, and the development of the mathematical techniques capable of revealing the interrelations between these various distributions can be performed at different levels of sophistication. A familiarity with the calculus of real variables, and with elementary statistical theory, however, is sufficient to yield a substantial number of interesting results and interrelationships which can be applied to many proposed mechanisms for cellular growth and division.

In order to establish the notation used in describing the statistical properties of microbial populations, some elementary results in probability theory are reviewed briefly in the next section. More extensive treatments may be found in various statistical texts (e.g. Cramer, 1955; Mood, 1950).

2.1. Notation and Ideas of Probability Theory

If the size classes of cells discussed in the previous section are made progressively smaller, then the histogram representing the probability of a randomly selected cell having its size in a particular range will tend towards a smooth curve. This size distribution is an example of a random variable having a continuous probability distribution. Representing this distribution by $f(s)$, the probability of a cell having its size (S) between two limits is given by (see Fig. 1)

$$\text{(probability that } S_1 < S < S_2) = \int_{S_1}^{S_2} f(s) \, ds$$

and $f(s)$ is assumed normalized, i.e.

$$\int_0^\infty f(s) \, ds = 1$$

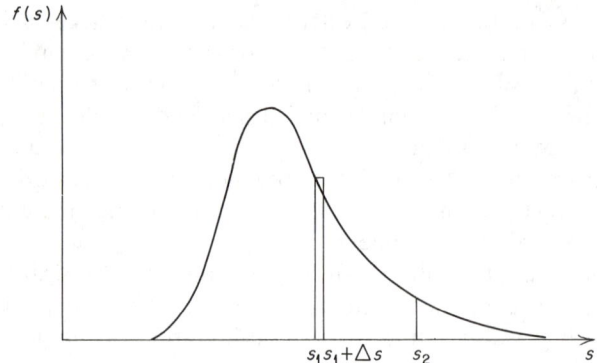

Fig. 1. A continuous probability distribution.

if s can only assume positive values. The expected value of any function of s is denoted by $E[h(s)]$ and is defined as

$$E[h(s)] = \int_0^\infty h(s)f(s)\,\mathrm{d}s$$

The generation-time distribution is the probability density function which yields the probability of interdivision times (τ) falling in a particular range, this variable is also limited to the range $0 < \tau < \infty$.

In addition to size and generation-time distributions it is possible to define (and in many cases to measure) various other distributions characteristic of the culture. These include, for example, the distribution of ages within the culture, the distribution of sizes for a given age, the distribution of generation rates (defined as the reciprocal of the age at division) etc. Some of these distributions are interrelated, and some are more interesting than others and have consequently received more attention. The interrelations between these distributions vary in mathematical complexity, and it is the intention in this chapter that these interrelations be developed for the most interesting cases, where the level of mathematical complexity allows.

Many mathematical functions can be used to represent probability distributions, and a most useful way of categorizing distributions is in terms of their moments. The moments of a distribution $f(x)$ of the continuous random variable x are defined as

$$\mu_p' = \int_0^\infty x^p f(x)\,\mathrm{d}x = E[x^p]$$

if $f(x)$ is non-zero for positive values of x only.

The first and second moments of the distribution specify the location and the dispersion of the distribution. The first moment μ'_1 is the mean of the distribution while moments about the mean are defined as

$$\mu_p = \int_0^\infty (x-m)^p f(x)\,dx = E[(x-m)^p] \qquad (1)$$

where

$$m = \int_0^\infty x f(x)\,dx = \mu'_1 \qquad (2)$$

e.g.

$$\mu_2 = \int_0^\infty (x^2 - 2xm + m^2) f(x)\,dx$$

$$= \mu'_2 - m^2 = \mathrm{var}(x) \qquad (3)$$

A parameter which is often used in connection with distributions of variables which assume only positive values, is the coefficient of variation ($c(x)$) which is a measure of the ratio of the spread (or dispersion) of the distribution to its mean, defined by:

$$c^2(x) = \frac{\mathrm{var}(x)}{m^2}$$

Under certain reasonable mathematical conditions it can be shown that a knowledge of all of the moments of a distribution specifies the distribution exactly, but in practice it is rarely possible or valuable to investigate the moments of the size and age distributions above the third moment. The third moment about the mean μ_3 gives a measure of the skewness of the distribution ($\mu_3 = 0$ for symmetrical curves). Since the size and age at division of a cell cannot generally be negative (except in a special sense which is occasionally used), and there is, in principle, no upper limit to these variables, the distributions may often be expected to be positively skewed (see, for example, Fig. 1).

It is often important to consider the joint distribution of two or more variables, for example, the joint distribution of age and size at division might be represented as $g(s, \tau)$ where

$$\int_0^\infty \int_0^\infty g(s, \tau)\,ds\,d\tau = 1$$

and the probability of a cell dividing with size between the limits s_1 and s_2 and age between t_1 and t_2 is given by

$$P(s_1 < s < s_2, t_1 < \tau < t_2) = \int_{s_1}^{s_1} \int_{t_1}^{t_1} g(s, \tau)\,ds\,d\tau \qquad (4)$$

The generation-time distribution ($f(\tau)$) can of course be obtained from this distribution since the probability of division at any age between t_1 and t_2 irrespective of the cells size is just given by

$$\int_{t_1}^{t_2} f(\tau) \, d\tau = \int_{t_1}^{t_2} \left[\int_0^\infty g(s, \tau) \, ds \right] d\tau$$

i.e. $f(\tau)$ is given by the marginal distribution $\int_0^\infty g(s, \tau) \, ds$. Moments for the bivariate distribution are defined similarly, clearly

$$\int_0^\infty \int_0^\infty \tau^2 g(s, \tau) \, ds \, d\tau = \int_0^\infty \tau^2 f(\tau) \, d\tau = \mu_2'(\tau) \tag{5}$$

with similar formulae for the moments of the distribution of size at division. A further set of joint moments can also be defined as the expected value of $[(\tau - \bar{\tau})^p (s - \bar{s})^q]$, the most important of these is the covariance

$$\text{cov}(s, \tau) = \int_0^\infty \int_0^\infty (\tau - \bar{\tau})(s - \bar{s}) g(s, \tau) \, ds \, d\tau \tag{6}$$

where $\bar{\tau}$ and $\bar{s}$ are the mean age and size at division. The condition that s and τ are quite independent (i.e. the probability of any value of one is unaffected by the value of the other) is equivalent to the condition that $g(s, \tau)$ factorizes into the product of functions dependent only on s and τ, i.e.

$$g(s, \tau) = g_1(s) g_2(\tau) \quad \text{for } s \text{ and } \tau \text{ independent} \tag{7}$$

In this case the covariance of s and τ is zero as can be seen immediately from eqn (6). The covariance gives a measure of linear dependence between τ and s. This dependence is generally measured by the correlation coefficient ρ, a normalized quantity defined by the expression:

$$\rho = \frac{\text{cov}(s, \tau)}{[\text{var}(s) \cdot \text{var}(\tau)]^{1/2}} \quad \text{where } -1 \leq \rho \leq 1 \tag{8}$$

For any two random variables having a joint distribution function, if $\rho = \pm 1$, then a complete linear dependence exists between them and either variable is a linear function of the other. If $\rho = 0$ the variables are said to be uncorrelated. It should be noted in passing, however, that while independent variables are always uncorrelated, uncorrelated variables are not necessarily independent (Cramer, 1955).

It is by no means certain that any correlation need exist between s and τ in the example chosen here. It is possible of course to define many other joint distributions of variables associated with the cellular division cycle. If, for example, we were to consider the joint distribution of the size of any cell and its age then a strong positive correlation would be expected. Ultimately, however, the only distributions worthy of serious study are those which can be related to a measurable distribution. Whereas the above-mentioned correlations could be measured by microscopic means, they cannot be measured directly by use of high-speed cell counters. Joint distributions of correlated variables where the correlation coefficient can be measured in this way are considered further in Section 5.

The case of independent random variables is particularly important in many applications, and use is often made of the properties of the distribution of combinations of independent random variables. In the analysis of the error in a calculated mathematical result formed from the sums and products of independent random variables each having an associated error for example, use is made of the simple relationships:

$$\text{var}(x_1 \pm x_2) = \text{var}(x_1) + \text{var}(x_2)$$

$$c^2(x_1 x_2) = c^2(x_1/x_2) = c^2(x_1) + c^2(x_2)$$

The first of these is an exact relationship while the second is an approximate result which is good for distributions having a small coefficient of variation.

2.2. Analytical Forms for Probability Distributions

Amongst the many mathematical functions which have been used to represent the probability distribution of a continuous random variable, one of the most important is the Gaussian or normal distribution given by

$$f(x) = \frac{1}{\sigma\sqrt{(2\pi)}} \exp\left\{-\frac{(x-m)^2}{2\sigma^2}\right\} \quad (9)$$

All values of x are allowed and the mean and variance of $f(x)$ are m and σ^2, respectively. The curve is symmetrical (in particular it has zero skewness) and is often considered inappropriate for age and size distributions since the probability of occurrence of negative values of these variables is not zero. However, for values of the coefficient of variation typically encountered ($c \simeq 0.2$), the probability of the normally distributed variable being less than zero is of the order of 10^{-6}. In view of the considerable algebraic simplification which often follows from the use of the normal distribution, extensive use will be made of this distribution in the following, when a specific form for a distribution is required or when relationships between distributions are considered. It will be assumed that we can always approximate an integral over a normal distribution (for example, when evaluating the expected value of some function of age or size) from zero to infinity by the integral over all values of the variable (positive and negative). A far more serious objection to the use of the normal distribution concerns its symmetry. While evidence on size distributions is inconclusive, it has often been claimed that the distribution of generation times, for example, is positively skewed, and cannot be adequately represented by a normal distribution. Recent data has shown, however, that for bacteria growing in agitated liquid culture, the generation-time distribution exhibits a zero skewness within the precision of the experiment over a large range of growth rates (Plank and Harvey, 1979), and is very well fitted by a normal distribution.

In many cases it is useful to make the approximation that one particular variable is essentially fixed in size. This corresponds to letting the variance in eqn (9) tend to zero. In this case the distribution becomes essentially a Dirac delta function

$$f(\tau) = \delta(\tau - \bar{\tau}) \tag{10}$$

This function is defined by the property

$$\int_{-\infty}^{\infty} f(x)\, \delta(x - a)\, \mathrm{d}x = f(a) \tag{11}$$

Other distributions which have been used in the literature include the Pearson type III distribution (see Section 7), the lognormal distribution, the reciprocal normal distribution and many others.

3. Steady-State Growth

For organisms growing by binary fission, the number of cells will double at regular intervals, if the growth rate is constant. Since there is generally a range of ages present in the culture, divisions occur at a steady (and increasing) rate. For cultures initiated with an inoculum of one (or a small number) of cells, the range of ages present at short times in the culture may differ from that obtained asymptotically if cellular growth and division were able to continue unabated indefinitely, and in this case the rate of increase of cell numbers may not initially be a smooth function of time. The natural variation in interdivision times occurring in a population will soon cause such fluctuations to die away, however, leaving the culture in what is generally called the exponential phase of growth since the number of divisions (and hence the rate of increase of cell numbers) is proportional to the number of cells in the culture, i.e.

$$\frac{\mathrm{d}N(t)}{\mathrm{d}t} = N'(t) = \nu N \tag{12}$$

1. MATHEMATICS OF MICROBIAL AGE AND SIZE DISTRIBUTIONS

where ν is a constant characteristic of the culture. The solution of eqn (12) is

$$N(t) = N_0 \exp[\nu(t - t_0)] \tag{13}$$

where N_0 is the number of cells at time t_0. For exponentially growing cultures, a plot of the logarithm of the number of cells ($\ln N(t)$) against time (t) will yield a straight line of slope ν.

In the steady state, the distribution of any variable associated with the cell must not depend on the time at which the sample is taken, and this may be taken as a definition of the steady state. In particular the volume distribution of cells must be constant with time. The combination of eqn (13) with this condition on the volume distribution acts as a constraint on other variables associated with the cell, since, for example, the rate of increase of total mass of some biochemical component must also be exponential with the same growth rate ν, otherwise the average amount per cell would increase or decrease indefinitely.

When a cell divides the total number of cells under consideration is increased by one cell. The two cells resulting from the division may be referred to as two daughter cells in which case the mother cell disappears, alternatively, one can consider that one daughter cell is formed and that the mother survives. In the following discussion the former viewpoint is used, thus new cells are formed at twice the rate at which cells divide, and the rate of increase of cell numbers in a culture is just the rate of division of mother cells

$$\text{division rate} = \nu N_0 \exp[\nu(t - t_0)] \tag{14}$$
$$= \tfrac{1}{2}(\text{rate of formation of new cells})$$

The rate of increase of the number of cells characterized according to any variable of the cell cycle must also increase according to eqn (13). It is important to realize that an exponential increase in the total mass of any cellular component does not imply an exponential increase in mass as a function of time for any individual cell. Since the exponential growth law applies only to the total quantity in the culture, it is sufficient that the amount per cell double in each cell cycle (i.e. in one generation time), the natural spread in ages of the cells will then yield exponential growth of the total quantity of this component.

As an example, it may be reasonable to expect that a cell which grows by extending in one dimension, while maintaining a constant cross-sectional area, will have a linear growth rate as a function of its age and will double its mass in one generation time. After division each daughter cell will also grow linearly but after n generations, the total mass of all progeny cells of the original cell will be 2^n times the original mass, thus linear growth for an individual cell still leads to exponential growth of the culture.

Another often used assumption, is that cell mass increases exponentially as a function of the age of the cell. Measuring the size of an individual cell or the mean size of a synchronous culture of cells as a function of the cell cycle may be expected to provide information on the growth law. It is extremely difficult, however, to distinguish between, for example, linear growth with a doubling in growth rate just prior to division, and exponential growth when cells are only observed over a size range of 2:1. In any case different growth laws may be expected for different organisms, and, in general, the mathematical development used here will not make specific assumptions about the rate of increase of mass of single cells as a function of time. When such assumptions are needed, a useful technique in the absence of good experimental data is to investigate the sensitivity of the predictions of any model to the assumed growth law.

4. Relationships between Distributions

By suitably designing an experiment, it is possible to measure the size distribution of, for example, all cells in a sample taken from an exponentially growing culture. It is also possible to measure the size distribution of newly formed cells using a technique which selects a sample of these cells. It is clear that these distributions will differ substantially, and need to be clearly distinguished. The nomenclature and notation used by authors in the past has varied, but Painter and Marr (1968) have suggested a standard scheme for naming the distributions which are most often encountered in studies of microbial population growth. This scheme will be followed here and is detailed in Table 1.

In addition to the symbols introduced in this table, the distribution of sizes (x) of extant growing cells is denoted by $\lambda(x)$, and the distribution of ages (a) of these cells by $U(a)$. In the remainder of this section various relationships between the distributions in this table will be developed.

Table 1. Probability distribution functions used to describe random variables associated with microbial populations.

Random variable	Newborn cells	Extant cells	Mother (dividing) cells
Size at birth	$\psi_n(x) \equiv \psi(x)$	$\psi_e(x)$	$\psi_m(x)$
Size at division	$\phi_n(x)$	$\phi_e(x)$	$\phi_m(x) \equiv \phi(x)$
Interdivision time	$f_n(\tau) \equiv f(\tau)$	$f_e(\tau)$	$f_m(\tau)$

4.1. Relationships between Generation-Time Distributions

It is clear that the distributions describing the size of cells which are newly formed and those which are dividing must differ, but it should be noted that Table 1 allocates different distributions for the generation time of extant cells and of newly formed cells. This difference is a little more subtle than that between size distributions, indeed no such distinction has as yet been made in this discussion, and at first sight it is only necessary to specify one distribution—that which specifies the probability of a particular generation (interdivision) time occurring in the population. The difference between these distributions arises, because, although we define an exponentially growing cell culture as a "steady-state" culture, the number of cells is not constant, and any sample selected from a growing population will contain more young organisms than old organisms. This in turn means that, dividing the culture into any two arbitrary groups, cells dividing at any instant are the sum of those which have short generation times and those with longer generation times. In the culture, however, there are more of the former and thus the probability of a particular generation time being exhibited by a dividing (or mother) cell is biased towards short generation times. In the preceding argument the cells have been divided into two groups, in reality an infinite variety of generation times are represented, and simple mathematical relationships exist between the three age distributions listed in Table 1. These relationships can be reached in different ways, and the derivation given below follows that of Painter and Marr (1968).

The number of cells in the culture with generation times less than some fixed generation time τ is, by definition

$$N_<(\tau) = N_0 \int_0^\tau f_e(z) \, dz \tag{15}$$

These cells will divide in the period from time zero to time τ, and at any instant (t) in this time interval, the rate of division within the culture is $\nu N_0 e^{\nu t}$ (i.e. the number dividing between t and $t + dt$ is $\nu N_0 e^{\nu t} \, dt$) (cf. eqn (14)). The cells dividing at this instant must be mother cells and the fraction of these with generation times greater than t (i.e. extant at time zero) and less than τ is given by $\int_t^\tau f_m(z) \, dz$. Thus another way of counting all cells in the culture with generation times less than τ is to sum over all divisions between 0 and τ

$$N_<(\tau) = \int_0^\tau \nu N_0 e^{\nu t} \int_t^\tau f_m(z) \, dz \, dt \tag{16}$$

Equating (15) and (16) gives

$$N_0 \int_0^\tau f_e(z)\, dz = N_0 \int_0^\tau \nu e^{\nu t} \int_t^\tau f_m(z)\, dz\, dt$$

changing the order of integration one can re-express the right-hand side of this equation as

$$N_<(\tau) = N_0 \nu \int_0^\tau f_m(z) \int_0^z e^{\nu t}\, dt\, dz$$

Differentiating now w.r.t. τ yields

$$f_e(\tau) = (e^{\nu\tau} - 1) f_m(\tau) \tag{17}$$

In a similar way the cells $N_<(\tau)$ can be counted by considering the rate at which they were formed in the past. All must have been formed in the period $-\tau$ to 0 since their generation times are less than τ and at any instant in this period the rate of formation of new cells was given by $2N_0\nu e^{\nu t}$ (eqn (14)), the fraction of these which survive until time zero and have generation times less than τ is given by $\int_{-t}^\tau f_n(x)\, dx$. Analogous manipulations to those above lead to the relationship

$$f_e(\tau) = 2(1 - e^{-\nu\tau}) f_n(\tau) \tag{18}$$

while equating the right-hand side of (17) and (18) yields

$$f_m(\tau) = 2 e^{-\nu\tau} f_n(\tau) \tag{19}$$

The distribution of generation times of newly divided cells is that which is intuitively closest to the distribution intended when one speaks loosely of the generation-time distribution within a culture, since it gives the probability of a brand new cell having a particular generation time. It is this distribution which is generally denoted by $f(\tau)$ without a subscript. The relationship (19) makes quantitative the statement above that the generation times of mother cells are biased towards shorter generation times, since the factor $e^{-\nu\tau}$ will emphasize the short generation-time contribution of $f(\tau)$ at the expense of the longer times. Clearly the extent of this distortion of $f(\tau)$ will depend upon the range of generation times which are exhibited; if $f(\tau)$ is non-zero over a very restricted range, then $2 e^{-\nu\tau}$ will be effectively constant (and in fact equal to 1, see eqn (22)) over this range and the distributions will be very similar.

These differences can be investigated quantitatively by making use of an approximate form for $f(\tau)$. Recent measurements with synchronous cultures have shown that $f(\tau)$ is very well represented by a Gaussian function of the form of eqn (9) with mean generation time $\bar{\tau}$. Using this form for $f(\tau)$ yields from eqn (19) by algebraic manipulation

$$f_m(\tau) = 2\exp[-\nu(\bar{\tau} - \tfrac{1}{2}\nu\sigma^2)]\frac{\exp[-(\tau - \tau_m)^2/2\sigma^2]}{\sigma\sqrt{(2\pi)}} \quad (20)$$

where $\tau_m = (\bar{\tau} - \nu\sigma^2)$, which indicates that $f_m(\tau)$ is also Gaussian with the same variance as $f(\tau)$ but with mean shifted by an amount given by $\nu\sigma^2$. We can also investigate the moments of $f_e(\tau)$ by writing (see eqn (18))

$$f_e(\tau) = 2f(\tau) - f_m(\tau)$$

Using the two equations for $f(\tau)$ and $f_m(\tau)$ we find that

$$\bar{\tau}_e = \int_0^\infty \tau f_e(\tau)\,d\tau = 2\bar{\tau} - \bar{\tau}_m$$

$$= \bar{\tau} + \nu\sigma^2$$

and

$$\overline{(\tau_e)^2} - (\bar{\tau}_e)^2 = \sigma^2(1 - 2\nu^2\sigma^2) \quad (21)$$

Thus the means $\bar{\tau}_e$ and $\bar{\tau}_m$ are symmetrically shifted a distance $\nu\sigma^2$ either side of $\bar{\tau}$ while $f_m(\tau)$ has the same variance as $f(\tau)$ and $f_e(\tau)$ has a variance which differs from that of $f(\tau)$ by $2\nu^2\sigma^4$.

It may readily be shown (see, e.g. Painter and Marr, 1968) that the function $f_m(\tau)$ is also a normalized distribution, i.e. that

$$\int_0^\infty f_m(\tau)\,d\tau = 1$$

Using eqn (20) we thus see that,

$$2\exp\{-\nu(\bar{\tau} - \tfrac{1}{2}\nu\sigma^2)\} = 1$$

since the Guassian component of $f_m(\tau)$ has an integral of unity (using the approximation of Section 2.2). Taking the logarithm of the above equation yields

$$\bar{\tau} = \ln(2)/\nu + \tfrac{1}{2}\nu\sigma^2 \quad (22)$$

a useful relationship between the mean generation time and the growth rate of the culture. As has been noted previously, the widely used assumption that $\bar{\tau} = \ln(2)/\nu$ is incorrect, but the inaccuracy involved is small for a typical spread in generation times. Equation (22) can also be derived under more general conditions (Painter and Marr, 1968).

We can now obtain an estimate of the differences between the three generation-time distributions discussed in this section. Using $\bar{\tau} \simeq \ln(2)/\nu$ and the fact that the coefficient of variation of the distribution $f(\tau)$ is remarkably constant for bacteria at a value close to 0.2 (Plank and Harvey, 1979) we find that $\nu\sigma^2 \approx 0.03\bar{\tau}$ and $2\nu^2\sigma^2 \approx 0.04$. This implies that the inaccuracy involved in the approximation $\bar{\tau} = \ln(2)/\nu$ is of the order of 1.5%. Furthermore the differences between the means of the distributions f_n, f_e and f_m are of the order of 3% and the variances differ by less than 4%. Taking into account the present precision in measurement of the moments of the generation-time distributions these differences have only academic significance. In the remainder of this article therefore, these small differences will be ignored and the phrase "generation-time distribution" will be assumed to apply to a single distribution characteristic of any sample taken from a microbiological culture. This simplification is, of course, justified only for distributions having a small coefficient of variation and growing by binary fission. In other situations due allowance may have to be made for the differences between the distributions.

Before leaving the subject of generation-time distributions, it is important to consider another aspect of the problem which calls into question the exact meaning of any distribution which is measured experimentally. The difference in the distributions f_n, f_e, f_m above was occasioned by the fact that an exponentially growing population of cells contains more cells with short generation times than cells with long generation times. The same effect ensures that newly divided cells will be the progeny of mothers which are biased with respect to their generation times. If the parent- and daughter-cell generation times are independent, then any bias with respect to the selection of newly divided cells is irrelevant to the discussion. The distinction between the generation-time distribution of newly divided organisms collected without bias with respect to their parent generation times and those collected from an exponentially growing culture was first pointed out by Powell (1964) who termed these distributions "artificial" and "real" distributions, respectively. In order to consider this effect mathematically, it is necessary to introduce the joint distribution of mother and daughter generation times (see Section 5). If this is defined such that the marginal distributions of mother and daughter generation times are alike equal to the generation-time distributions, then the single and bivariate distributions so defined are called "artificial" since mother and daughter cells of the same generation time do not occur with equal frequency in a batch culture. The "real" distribution of generation times is the distribution of generation times belonging to newly divided cells from an exponentially growing population. It is similarly possible to define a "real" joint distribution of parent and daughter generation times. These distributions are related, and in certain circumstances relations between them can be formulated as integral equations. If, of course, the parent- and daughter-cell generation times are independent then the difference disappears. This distinction is not dealt with any further here in spite of the evidence showing strong correlations between parent- and daughter-cell generation times (Plank and Harvey, 1979) in batch culture since the distinctions have been shown to lead to differences between the distributions of the same order as those found above for the functions f_n, f_e and f_m. Further discussion of these points will be found elsewhere (see, e.g. Powell, 1969; Painter, 1975; and references therein).

4.2. Relationships between the Size Distributions

Since each dividing cell belongs to the class characterized by the distribution ϕ, and produces two daughter cells characterized by the distribution ψ, it is to be expected that a mathematical relationship exists between them. Clearly if division is symmetrical with each mother cell producing two daughter cells of exactly half the size of the mother then

$$\psi(x) = 2\phi(2x) \qquad (23)$$

If division is not symmetrical then it is convenient to introduce a distribution giving the frequency of a daughter cell of size x for a mother cell of size x_m. If each mother cell of size x_m produces a daughter cell of size x to $x + dx$ with probability $K(x_m, x)\,dx$ then daughter cells of this size will be represented with probability

$$\psi(x) = \int_x^\infty \phi(x_m) K(x_m, x)\,dx_m \qquad (24)$$

since daughter cells must be smaller than mother cells.

This equation is of little value in the absence of further information concerning the function K. It is clear that it must be a function of x which is symmetrical about $x_{m/2}$, but some other assumption needs to be made concerning the form of the function as x_m varies. Koch and Schaecter (1962) introduced the idea later extended by Powell (1964) of using as a random variable the ratio of the size of the daughter to that of the mother and assuming that this distribution was independent of the size of the mother. Following Painter and Marr (1968) we introduce a distribution $k(r)$ for the random variable $r = x/x_m$. Since the probability of a daughter cell in the size range 0 to x_0 is the probability of a daughter cell having r in the range 0 to x_0/x_m then

$$\int_0^{x_0} K(x_m, x)\,dx = \int_0^{x_0/x_m} k(r)\,dr$$

By differentiation

$$x_m K(x_m, x_0) = k(x_0/x_m)$$

and substituting in eqn (24) gives

$$\psi(x) = \int_x^\infty \phi(x_m) \frac{k(x/x_m)}{x_m} \, dx_m \qquad (25)$$

which can also be written as

$$\psi(x) = \int_0^1 \phi(x/r) \frac{k(r)}{r} \, dr \qquad (26)$$

If k is represented by a Dirac delta function then eqn (25) reduces to eqn (23) of course. Microscopic evidence indicates that the function k can be well represented by a Gaussian function with a very small coefficient of variation, for bacteria, and the assumption of equal sizes of daughter cells is a good approximation for these cells (see, e.g. Harvey et al., 1967).

If it is assumed that all cells within the culture remain viable then it is possible in principle to keep track of each cell, that is, a cell either grows from one size to a larger size or it divides forming two new smaller cells. This "conservation law" acts as a constraint on the size distributions in Table 1, and leads to an equation first derived by Collins and Richmond (1962). This equation follows from a consideration of the fate of cells having a size less than x. Using the definition of the function λ the number of these is given by

$$N \int_0^x \lambda(z) \, dz$$

The rate of increase of cells in this size range is just ν times this quantity and this rate of increase can be accounted for as the sum of the net rate at which cells are added and removed by division and the rate at which they are lost from this size class by growth to a larger size than x. The rate of change of cell numbers by birth and divisions is given by (see eqn (14))

$$2kN \int_0^x \psi(z) \, dz - kN \int_0^x \phi(z) \, dz$$

while the rate of loss by growth depends on the specific growth rate $V(x)$; in a short time dt, cells in the size range $x - dx$ to x will grow out of this size class (where $dx = V(x) \, dt$), there are $N\lambda(x) \, dx$ of these cells thus the rate of loss is $-N\lambda(x) \, dx/dt$, or $-N\lambda(x)V(x)$. A small fraction of cells in the size range x to $x - dx$ will also divide or be new born in time dt. This fraction gives a second-order correction to the rate of loss of cells from this size class which can be neglected as dx tends to zero. The net rate of growth of the number of cells in the size range 0 to x is thus

$$\nu N \int_0^x \lambda(z) \, dz = kN \left\{ 2 \int_0^x \psi(z) \, dz - \int_0^x \phi(z) \, dz \right\} - N\lambda(x) V(x)$$

or

$$V(x) = \frac{\nu}{\lambda(x)} \int_0^x [2\psi(z) - \phi(z) - \lambda(z)] \, dz \tag{27}$$

which is the equation of Collins and Richmond (1962).

4.3. Relationships between Size and Generation-Time Distributions

Since it is clear that the size and the age of cells are strongly correlated, it is possible that the distribution of generation times may be defined by the distribution of sizes at division of mother cells and the growth rate of the cells. While connections between the distribution of sizes and generation times are generally formulated in terms of the specific dictates of a particular model, it is possible to formulate a relatively general relationship based on certain reasonable assumptions. The following probabilistic arguments are due to Eakman (1966) and are restricted only by the assumption that the growth rate of a cell depends only on its mass (and thus implicitly upon its age) and not on absolute time, i.e. that the function

$$V(x) = \frac{dx}{dt} = V\{x(x_0, t)\} \tag{28}$$

where the notation indicates that the present size x is an implicit function of the size (x_0) when newly divided and the age of the cell (t).

The probability that a cell will have generation time in the range τ to $\tau + d\tau$ is given by

$$P(\tau) = f(\tau) \, d\tau$$

The probability that a cell has generation time in the range τ to $\tau + d\tau$ given that its initial size was in the range x_0 to $x_0 + dx_0$ is given by $\phi_n(x) \, dx$ (from the distribution of sizes of newly formed cells at the time of their subsequent division). This conditional probability $P\{\tau|x_0\}$ is thus given by

$$P\{\tau|x_0\} = \phi_n\{x(x_0, \tau)\} V\{x(x_0, \tau)\} \, d\tau \tag{29}$$

Thus the probability of any cell having this generation time given that its parent had mass x_m at fission and that it had an initial mass x_0 is $P\{\tau|x_0\}$ times the probability of such a size at birth, i.e.

$$P\{\tau|x_0 x_m\} = \phi_n(x_0, \tau) V(x_0, \tau) K(x_m, x_0) \, dx_0 \, d\tau$$

In order to obtain the total probability of such a generation time for a given parental mass x_m we need to sum over all initial masses x_0:

$$P\{\tau|x_m\} = d\tau \int_0^{x_m} \phi_n(x_0, \tau) V(x_0, \tau) K(x_m, x_0) \, dx_0$$

The probability that a parent cell will have a mass in the range x_m to $x_m + dx_m$ is, however, given by $\phi(x_m) \, dx_m$ thus we can finally obtain the total probability of a cellular generation time in the range τ to $\tau + d\tau$ by summing over all possible masses of the mother cell

$$P\{\tau\} = \int_0^\infty \phi(x_m) P\{\tau|x_m\} \, dx_m$$

or

$$f(\tau) = \int_0^\infty \phi(x_m) \, dx_m \int_0^{x_m} \phi_n(x_0, \tau) V(x_0, \tau) K(x_m, x_0) \, dx_0 \quad (30)$$

Alternatively we can use eqn (29) together with the probability of a particular size at birth $\psi(x_0) \, dx_0$, and sum over the sizes x_0 to obtain the form:

$$f(\tau) = \int_0^\infty P\{\tau|x_0\} \psi(x_0) \, dx_0$$

$$= \int_0^\infty \phi_n(x_0, \tau) V(x_0, \tau) \psi(x_0) \, dx_0 \quad (31)$$

It is important to note that eqns (30) and (31) both contain the distribution of sizes at division of a sample of newly formed cells $\phi_n(x)$ and it is not correct to identify this with the distribution $\phi(x)$ $(=\phi_m(x))$.

The difference between the functions $\phi_n(x)$ and $\phi(x)$ is due to the differing age distribution of cells of a particular size in each population. Since cell growth is a monotonic function of age, smaller cells tend to have smaller ages and thus small dividing cells were born at a time when the rate of addition of daughter cells was higher than the rate existing when larger dividing cells were born. The distribution $\phi_n(x)$ corresponds to the probability of division of a cell at a particular size when all dividing cells were added at the same rate (since they were born at the same time). These distortions or differences between the ϕ_n and ϕ_m distributions are of the same order as the differences between the corresponding generation-time distributions and can also be ignored when the precision with which these distributions can be experimentally measured is considered. This can be seen by considering the time over which presently dividing cells in a culture had previously been born. If this time is short compared to the mean generation time, then the above approximation will be a good one. Since this range in times is the range of the function $f_m(\tau)$, the corrections to the mean and variance will be of the same order as those found previously for the generation-time distributions. In this approximation then, eqn (30) can be written

$$f(\tau) = \int_0^\infty \phi(x_m) \, dx_m \int_0^{x_m} \phi(x_0, \tau) V(x_0, \tau) K(x_m, x_0) \, dx_0 \qquad (32)$$

a form originally quoted by Eakman et al. (1966). If we make the approximation that all daughter cells are exactly half the size of the mother cells then

$$K(x_m, x) = \delta(x - \tfrac{1}{2}x_m)$$

and

$$f(\tau) = \int_0^\infty \phi(x_m) \phi(\tfrac{1}{2}x_m, \tau) V(\tfrac{1}{2}x_m, \tau) \, dx_m \qquad (33)$$

This equation permits the calculation of $f(\tau)$ from the distribution of sizes of dividing cells (or from the distribution $\psi(x)$ using eqn (31)) for an assumed growth law. Since some of the more successful models of the cellular division process involve primary control of division by the size of the cell, an application of the above equation is considered in Section 7.

4.4. The Age Distribution

It has already been noted that in an exponentially growing culture of cells, younger cells are more numerous than older cells. The precise expression of this statement is contained in the distribution of ages of cells in the culture. This distribution (i.e. that of the elapsed time since the most recent cell division) is of some importance since it is often assumed that the synthesis of some biochemical component of the cell proceeds at a known rate as a function of time after initiation, and that this initiation occurs at a time which is precisely related to the time of cellular division. The distribution of mass of this component per cell will then be determined by the distribution of cell ages in the population.

This distribution is readily obtained by considering the rate at which cells of a particular age were formed. Cells of age a were formed at a time $t = -a$ and at this time the rate of formation of new cells in the culture was (see eqn (14)) $2N_0 \nu e^{-\nu a}$, thus the number formed between $t = -a$ and $t = -a - da$ was $2\nu N_0 e^{-\nu a} da$. The probability of a cell newly divided at this time reaching time $t = 0$ without dividing is given by {1 − probability of division before time a} thus the total number of cells of age a to $a + da$ is given by

$$U(a)\,da = 2\nu N_0 e^{-\nu a}\left\{1 - \int_0^a f(t)\,dt\right\}da \tag{34}$$

since the generation-time distribution is normalized.

In the event that the generation-time distribution has a very small coefficient of variation, we can approximate $f(t)$ by the Dirac delta function (eqn (10)) to obtain

$$U(a) = \begin{cases} 2\nu e^{-\nu a}, & a < \bar{t} \\ 0, & a > \bar{t} \end{cases}$$

5. Correlations between Variables Associated with the Cell Cycle

As has already been mentioned, several of the variables associated with the cell may be expected to be correlated. The age of a cell, for example, is correlated with the cell's size. Microscopic observations (see Section 6) have produced evidence for a positive correlation between sister-cell generation times and for a negative correlation between mother- and daughter-cell generation times (i.e. long-lived mother cells tend to produce short-lived daughter cells and vice versa).

Mathematically, these correlations are a consequence of the form of the bivariate distribution of the random variables involved. Consider the case of the correlations between the sizes of sister cells. If the bivariate distribution of sizes is denoted by $L(x_1, x_2)$, where x_1 and x_2 are the sizes of the two sister cells, then this distribution provides an alternative description to that of the previous section of the probability of a newly formed cell having its size in a particular range, in particular

$$\psi(x) = \int_0^\infty L(x, z)\,dz = \int_0^\infty L(z, x)\,dz$$

A small spread in the sizes of daughter cells for a given size of mother cell has been accounted for in Section 4 by introducing the distribution $k(r)$. No matter what the form of $L(x_1, x_2)$ we must have the sum of the

1. MATHEMATICS OF MICROBIAL AGE AND SIZE DISTRIBUTIONS

sizes of the daughter cells equal to the size of the mother cell, thus $(x_1 + x_2)$ is a random variable with the distribution $\phi(x_m)$. Clearly, the expected value of $(x_1 + x_2)$ is $\bar{x}_m$, the mean size of dividing cells, and the variance of $\phi(x)$ is given by

$$\text{var}[\phi(x)] = E[(x_1 + x_2)^2 - \bar{x}_m^2]$$

$$= \int_0^\infty \int_0^\infty [(x_1 + x_2)^2 - \bar{x}_m^2] L(x_1, x_2) \, dx_1 \, dx_2$$

$$= 2 \int_0^\infty (x - \tfrac{1}{2}\bar{x}^m)^2 \psi(x) \, dx$$

$$+ 2 \int_0^\infty \int_0^\infty (x_1 - \tfrac{1}{2}\bar{x}_m)(x_2 - \tfrac{1}{2}\bar{x}_m) L(x_1, x_2) \, dx_1 \, dx_2$$

$$= 2 \, \text{var}[\psi(x)][1 + \rho(x_1, x_2)] \tag{35}$$

Where the mean size of daughter cells is $\tfrac{1}{2}\bar{x}_m$ and $\rho(x_1, x_2)$ is the correlation coefficient between the sizes of the daughter cells defined in eqn (8). In principle, a measurement of the variance of $\phi(x)$ and of the variance of $\psi(x)$ yields $\rho(x_1, x_2)$ from eqn (35). Direct microscopic measurement will yield estimates of $\psi(x)$ and $\phi(x)$ as well as $\rho(x_1, x_2)$ but whereas $\psi(x)$ and $\lambda(x)$ are accessible using an automatic cell counter, the distribution of sizes of dividing cells is not and $\rho(x_1, x_2)$ cannot thus be derived from such measurements. It may be suggested that the distribution $\phi(x)$ could be determined from the Collins–Richmond equation using an assumed form for the specific growth rate (for example, exponential growth). Unfortunately, use of this equation with reasonable estimates of the distribution $\phi(x)$ have shown that $V(x)$ is not realistically represented by a simple exponential or linear function (Harvey et al., 1967). Alternatively microscopic measurement of $\rho(x_1, x_2)$ could provide a means of obtaining the second moment of the distribution $\phi(x)$ from that of $\psi(x)$ and the Collins–Richmond equation could then be used to obtain the specific growth rate as a function of cell size. It should be noted that, because of the limits on $\rho(x_1, x_2)$, the variance of $\phi(x)$ must lie between zero and four times that of $\psi(x)$. The case of each daughter cell having the same size (half that of the mother) corresponds to $\rho(x_1, x_2) = 1$ and the variance of $\phi(x)$ is then four times that of $\psi(x)$ as may be verified by direct computation of the variance of the distribution given in eqn (23). Other important joint distributions of correlated variables include the distribution of mother–daughter generation times and that of sister–sister generation times. Of these, the former plays an important role in determining the rate of growth of a synchronous culture of cells, and is considered further in the next section.

5.1. The Development of Synchronous Cultures

Attempts have been made to synchronize the divisions of the cells in a culture by various means for many years. The advantages of synchronized growth lie in the potential for studying key biochemical processes as a function of the age of the cell. Consider a culture of cells selected in some way such that each cell has just divided in a time prior to $t = 0$ which is negligible compared with the generation time of the cells. These cells will have the size distribution $\psi(x)$, and the probability of cell division will be very small until cells reach an age which becomes comparable with the mean generation time (assuming that the distribution has a small coefficient of variation). The total number of cells in the culture will thus be a constant and will then double over a period near to $\bar{t}$. The successive divisions of the progeny cells give rise to a series of steps in a graph plotting the number of cells against time. In order to describe the growth of such cultures mathematically, it is convenient to define t_j, a random variable representing the generation time of a jth generation organism, and to define

$$T_j = t_1 + t_2 + \ldots + t_j$$

T_j is a random variable representing the time of division of a jth generation organism in the synchronous culture. If T_j has the distribution $G_j(T_j)$ then the probability of a jth generation organism dividing before time T is given by $\int_0^T G_j(t)\,dt$. Thus the number of jth generation organisms at any time is given by the difference between the number formed at this time (or the number collected initially for $j = 1$) and the number that have divided to produce $j + 1$th generation cells, i.e.

$$N_1(T) = N_0\left(1 - \int_0^T G_1(t)\,dt\right)$$

and

$$N_j(T) = 2^{j-1}N_0\left(\int_0^T G_{j-1}(t)\,dt - \int_0^T G_j(t)\,dt\right)$$

Thus the total number of cells at any time in the culture will be given by

$$N(T) = \sum_{j=1}^{\infty} N_j(T)$$

$$= N_0\left(1 + \sum_{j=1}^{\infty} 2^{j-1}\int_0^T G_j(t)\,dt\right)$$

1. MATHEMATICS OF MICROBIAL AGE AND SIZE DISTRIBUTIONS 23

and

$$N'(T) = 2^{j-1} N_0 \sum_{j=1}^{\infty} G_j(T) \qquad (36)$$

If it is assumed that the lifetimes of different generations of cells are independent then the functions $G_j(t)$ become the distribution of the sum of j independent random variables each with the same distribution $(f(t))$. The properties of such distributions are well known (see, e.g. Cramer, 1946), in particular $G_j(t)$ becomes the j-fold convolution of the generation-time distribution and each term in eqn (36) becomes the convolution of the generation-time distribution with the previous term

$$N'(T) = N_0 \left\{ f(T) + 2 \int_0^T f(t) f(T-t) \, dt + \cdots \right\} \qquad (37)$$

which may be written

$$N'(T) = N_0 f(T) + 2 \int_0^T N'(T-t) f(t) \, dt \qquad (38)$$

This equation was elegantly derived by Harris (1959) from renewal theory, and has been used extensively in the literature concerning synchronous cultures. The properties of the distribution of a sum of independent random variables lead to the following results for the first few moments of the functions $G_j(T)$

$$\int_0^\infty G_j(T) \, dT = 1; \quad \int_0^\infty T G_j(T) \, dT = j\bar{\tau}$$

$$\int_0^\infty (T - j\bar{\tau})^2 G_j(T) \, dT = j\sigma^2; \quad \int_0^\infty (T - j\bar{\tau})^3 G_j(T) \, dT = j\mu_3 \qquad (39)$$

where σ^2 and μ_3 are the variance and third central moment of $f(t)$. Thus the graph of $N'(T)$ becomes a sum of the functions $G_j(T)$ which form a series of equally spaced distributions along the time axis. Strong correlations between parent and daughter generation times lead to serious disagreement between experiment and the predictions of eqn (38) when T becomes comparable with $2\bar{\tau}$. In particular the variance of $G_2(T)$ is less than twice the variance of $G_1(T)$ and the correlations must be accounted for in the formalism. Introducing the joint distribution $H(t_1, t_2)$ of parent and daughter generation times we must have (neglecting the small differences in generation-time distributions discussed at the end of Section 4.1)

$$f(t_1) = \int_0^\infty H(t_1, t_2) \, dt_2 \quad \text{and} \quad f(t_2) = \int_0^\infty H(t_1, t_2) \, dt_1$$

The mean of $G_2(T)$ is now given by

$$E(t_1 + t_2) = \int_0^\infty T G_2(T) \, dT$$

$$= \int_0^\infty \int_0^\infty (t_1 + t_2) H(t_1, t_2) \, dt_1 \, dt_2$$

$$= 2\bar{\tau}$$

Similarly the variance of $G_2(T)$ is given by (compare eqn (35))

$$\text{var}(G_2(T)) = 2\sigma^2(1 + \rho^{(1)}) \tag{40}$$

where

$$\sigma^2 \rho^{(1)} = \int_0^\infty \int_0^\infty (t_1 - \bar{\tau})(t_2 - \bar{\tau}) H(t_1, t_2) \, dt_1 \, dt_2$$

We note that variations in $\rho^{(1)}$ can change the variance of $G_2(T)$ from twice that given by the convolution of $f(t)$ with itself ($\rho^{(1)} = 1$), to zero ($\rho^{(1)} = -1$). The condition $\rho^{(1)} = -1$ corresponds to each daughter cell having a generation time of exactly $(2\tau - t_2)$, where t_2 was the generation time of the mother, and in this case all second generation cells divide at the same time.

The results given here can be extended to yield formulae for the moments of $G_j(T)$ for arbitrary j (Harvey, 1972a), the result, for example, for the variance of $G_3(T)$ is

$$\text{var}(G_3(T)) = 3\sigma^2 + 2\sigma^2(2\rho^{(1)} - \rho^{(2)}) \tag{41}$$

where $\rho^{(2)}$ is the mother–granddaughter generation-time correlation coefficient.

We have so far considered only the time development of the number of cells in the synchronous culture which yields information on the generation-time distribution and on the correlation coefficients $\rho^{(n)}$. The size distribution considered in Section 4 referred to cells in balanced exponential growth and, for example, the Collins–Richmond equation cannot be used to describe the growth of synchronous cultures. A modified version of this equation can be developed for synchronous cultures (Painter and Marr, 1968) but the equation becomes algebraically complicated and the interested reader is referred to the original paper.

1. MATHEMATICS OF MICROBIAL AGE AND SIZE DISTRIBUTIONS 25

Instead of considering the size distribution of cells in a synchronous culture as a function of time, it is simpler to consider the average cell volume as a function of time. Sinclair and Ross (1969) developed a theory of cell growth and division in synchronous populations under the assumption that control of division was independent of volume, and compared linear and exponential modes of volume increase. Their treatment was extended by Trucco (1970) wherein the mean cell volume of a synchronous population was calculated from an age-dependent branching process. Both of these treatments implicitly or explicitly make use of the renewal eqn (38) which applies only to cultures in which there is no dependence between parent and progeny cell generation times. In view of the strong correlations which often exist between mother- and daughter-cell generation times, it is preferable not to rely on the renewal equation. Using the normal form for the generation-time distribution, and the general results obtained above, it is relatively simple to deduce an expression for the mean cell volume of a particular generation of cells in a synchronous culture.

At any time (t) after the synchronous culture is initiated, the size of the cells within the culture will be given (using the approach of Section 4.3) by:

$$x_1 = \alpha(x, t) = \int_0^t V(x, t) \, dt$$

where x was the initial mass of the cell and α is some function of these variables. In the case of exponential growth, for example, $x_1 = x \, e^{\nu t}$. The size of a second-generation cell whose mother divided at time t_1 is then given by

$$x_2 = \alpha[\tfrac{1}{2}\alpha(x, t_1), (t - t_1)]$$

Since the cells in the culture divide at various times, the mean size of second-generation organisms as a function of time is given by

$$\bar{x}_2(t) = \frac{\int_0^t \alpha[\tfrac{1}{2}\alpha(x, t_1), (t - t_1)] G_1(t_1) \, dt_1}{\int_0^t G_1(t_1) \, dt_1} \qquad (42)$$

This process is readily extended to higher generations and the mean size can be evaluated from equations analogous to that above. In the simple example of exponential growth, eqn (42) reduces to the form

$$\bar{x}_2 = \tfrac{1}{2} x \, e^{\nu t}$$

Different models for the growth of individual cells can lead to readily discernible differences in the mean cell volume of a particular generation of cells as predicted, for example, by eqn (42) for second-generation cells. Since different generations of cells can often be distinguished (if, for example, the distributions $\psi(x)$ and $\phi(x)$ do not overlap), this approach is useful in the analysis of size distributions encountered in synchronous cultures.

6. Experimental Results for Size and Age Distributions

Two main techniques have yielded information on the size and generation-time distribution of a culture of cells, and on the dependence of these distributions upon environmental parameters. Extensive measurements of the generation-time distribution have been made by Powell (see Powell and Errington, 1963, and references therein) using a carefully designed culture chamber and microscopic observation. Powell's measurements indicate that the generation-time distribution $f(\tau)$ is a unimodal, positively skewed function. Later measurements by Kubitschek (1962, 1966), also using microscopic observation, have supported the hypothesis that $f(\tau)$ is a positively skewed function. It is, however, very difficult to maintain organisms in a state of balanced growth in the restricted environment necessary for microscopic observation, and it has been pointed out (Koch and Schaecter, 1962) that the skewness of the distribution has decreased as experimental techniques have improved.

The generation-time distribution was first extracted from synchronous culture data by Marr *et al.* (1969) using the assumption of independence of mother–daughter generation times and under more general circumstances by Harvey (1972). As discussed in Section 5 the generation-time distribution and the parent–daughter generation-time correlation coefficient can be derived from the observed rate of growth of a synchronous culture. The problems associated with this technique lie not in the statistical precision of the measurement of the number of cells but in the data analysis of the recorded cell numbers. Although the concentration of cells in the culture is a smooth function of time, the measurement of this concentration cannot be exact, and at best yields a series of points each with an associated experimental error. Direct numerical differentiation of the points will yield a series of values which will be greatly influenced by the experimental errors. The most satisfactory method for dealing with the data is to "smooth" it by digital filtering. In this technique the data points are convoluted with the Fourier transform of a step function. The resulting data can then have no Fourier components above a predetermined cutoff frequency (Harvey, 1972b). The resulting smoothed data is then differentiated to yield the functions $G_j(t)$. Early measurements using data from a culture synchronized using the membrane elution technique yielded a generation-time distribution which exhibited no detectable skewness. More recent measurements on synchronous cultures prepared by velocity sedimentation (i.e. by selection of the smallest cells) have shown this to be true over a wide range of growth rates (Plank and Harvey, 1979) for cultures of *Escherichia coli* B/r in a state of balanced growth. In all cases the coefficient of variation of the distribution was close to 0.2 and the distribution was indistinguishable within experimental error from a Gaussian distribution.

In using synchronous cultures to determine the parameters of the generation-time distribution, care must be taken to ensure that the balanced growth of the organism is disturbed as little as possible by the selection procedure. In this connection it should be noted that a culture reflecting balanced growth must show a gradual loss of synchrony due to the natural spread of generation times within the culture. An over zealous pursuit of synchrony can lead to cultures which are far from the steady state of growth. Synchronous cultures induced by dilution from the early stationary phase, for example, often show abnormally sharp steps in concentration. Whilst these cultures are of some interest, they do not represent the normal state of exponential growth in the first few generations of multiplication.

The advent of modern cell counters has also provided a means for measuring the distribution of sizes of cells. In order for the measurement to provide a useful estimate of the size distribution, much care must be taken with the design of the counter (see, e.g. Kubitschek, 1969) although it is usually possible to obtain a reliable estimate of the mean size of a sample. The number of cells in the sample, of course, can be accurately determined under controlled conditions.

Microscopic observation of the generation time of individual organisms permits the calculation of mother–daughter and sister–sister generation-time correlation coefficients. Powell (1958) reported strong positive correlations between sister generation times and little correlation between mother–daughter generation times. Negative correlations between mother- and daughter-cell generation times were reported by Kubitschek (1966) but the coefficients were not determined accurately. Synchronous cultures have shown that the mother–daughter generation-time correlation coefficient is close to -0.5 over a range of growth rates with doubling times of less than 65 min (Plank and Harvey, 1979). Models of the cellular division process must be capable of explaining these features of the distribution of generation times and the measured correlation coefficients. Some specific models are considered in the next section.

7. Specific Models for Cell Growth and Division

Rahn (1932) and Kendall (1948) proposed two early models for the control of division which are illustrative of the general method of proposing a model and the development of its mathematical consequences. Rahn proposed that division occurred as soon as some fixed number (g) of events had occurred inside the cell, and that these events were independent (i.e. could occur simultaneously). Kendall proposed instead that these g events were constrained to occur in a given order. If it is supposed that the rate of occurrence of the next event is β, and the probability that n events have occurred by age τ is denoted as $P_n(\tau)$, then it is straightforward to derive a system of equations relating the probabilities $P_n(\tau)$ using Kendall's model.

For $n > 1$, exactly n events can occur in the time $(0, \tau + \Delta\tau)$ either by having $n - 1$ events in the interval $(0, \tau)$ and 1 in $\Delta\tau$ or n events in the interval $(0, \tau)$ and none in $\Delta\tau$ (ignoring the second-order terms involving more than one event in $\Delta\tau$ which become insignificant as $\Delta\tau \to 0$), thus

$$P_n(\tau + \Delta\tau) = P_n(\tau)(1 - \beta\Delta\tau) + P_{n-1}(\tau)\beta\Delta\tau$$

or using the definition of the derivative

$$\frac{dP_n(\tau)}{d\tau} = -\beta P_n(\tau) + \beta P_{n-1}(\tau)$$

By similar reasoning the special case $n = 0$ yields

$$\frac{dP_0(\tau)}{d\tau} = -P_0(\tau)$$

subject to the initial conditions

$$\left.\begin{array}{l} P_n(0) = 0 \\ P_0(0) = 1 \end{array}\right\} \quad n > 1$$

The solution to this system of equations is the Poisson distribution

$$P_n(\tau) = e^{-\beta\tau}(\beta\tau)^n/n! \tag{43}$$

as may be verified by substitution and can readily be reached by considering the Laplace transformation of the above equations.

The expected rate at which the gth event occurs in the time interval τ to $\tau + \Delta\tau$ is $\beta P_{g-1}(\tau)\Delta\tau$ and this is (by the postulate of the model) also given by $f(\tau)\Delta\tau$. Thus

$$f(\tau) = e^{-\beta\tau}(\beta\tau)^{g-1}/(g-1)! \tag{44}$$

This is a Pearson type III distribution. The hypothesis of Rahn leads to a different distribution given by

$$f(\tau) = g\beta e^{-\beta\tau}(1 - e^{-\beta\tau})^{g-1} \qquad (45)$$

The fact that the hypothesis of Kendall leads to a Pearson type III distribution means that the coefficient of variation of the distribution has a special significance

$$c^2 = 1/g$$

and thus the constancy of the coefficient of variation for a range of growth rates in a particular organism is readily accounted for: it is simply a consequence of the fact that the same number of primitive events (for example, synthesis of components of the cell) have to be completed.

As Powell (1958) has pointed out, the variation of c^2 with increasing complexity of growth media is of some interest, since a steadily increasing value of c^2 with increasing complexity of the medium (i.e. in a situation where fewer components have to be synthesized) would constitute support for a model of Kendall's type. In the most recent experiments using synchronous cultures, however, no such trend was found.

A further inadequacy of the type of model proposed above is that it fails to account for the observed correlations between mother and daughter and sibling generation-time correlation coefficients. Koch and Schaechter (1962) were the first to propose a model which accounted for these correlations in a simple fashion. The approach used in this work was to assume that cell division occurred when the cell size reached a critical value. This model and its consequences have been developed in detail by Powell (1964) and others, and when investigated rigorously lead to mathematical relationships of some complexity. The basic ideas of the model have been very useful, however, in the development of postulated mechanisms for the control of cell division and, in spite of criticisms of the details of the model, its predictions are still in good agreement with experimental data on the generation-time distribution and the mother–daughter generation-time correlation coefficient. Recent measurements on the rate of increase of cell length have lent support to this simple model, since evidence has been found that some bacterial cells divide a short time after the attainment of a critical length (this evidence is reviewed by Sargent (1978)). Noting that the coefficient of variation of the size at division was much less than that of the distribution of generation times, Koch and Schaechter proposed that cellular division was controlled more strongly by cell size than cell age. Some consequences of this proposal are yielded readily from results obtained in previous sections. Assuming that the rate of growth of a single cell is exponential and that division results in daughter cells of equal size, we see from eqn (33) that (dropping the subscript on x_m) if $V(x, \tau) = \nu x\, e^{\nu \tau}$

$$f(\tau) = \tfrac{1}{2}\nu e^{\nu \tau} \int_0^\infty x\phi(x)\, \phi(\tfrac{1}{2}x\, e^{\nu \tau})\, dx \qquad (46)$$

As Powell (1964) has pointed out, the substitution $y = 2^{-1/2} x \exp(\tfrac{1}{2}\nu t)$ yields

$$f(\tau) = \nu \int_0^\infty y\phi\{\exp[-y\exp(\xi)]\}\phi\{\exp[y\exp(\xi)]\}\, dy$$

where

$$\xi = \tfrac{1}{2}(\nu \tau - \ln(2))$$

thus $f(\tau)$ is symmetrical about $\tau = \log(2)/\nu$ and the model predicts no skewness of the distribution. It is interesting to investigate the form of this distribution for some specific postulate for the shape of the size distribution ϕ. An alternative approach (Plank, 1978) is to investigate the moments of $f(\tau)$ in terms of the moments of ϕ, to obtain relationships which are independent of the form of ϕ.

If we use a Gaussian form for ϕ

$$\phi(x) = \left(\frac{1}{\sigma\sqrt{(2\pi)}}\right) \exp[-(x - \bar{x})^2/2\sigma^2]$$

then integration of eqn (46) leads to the following form for $f(\tau)$ (again using the good approximation discussed in Section 2 since the coefficient of variation of ϕ is typically of the order of 0.1)

$$f(\tau) = \left(\frac{\nu \bar{x}}{\sigma\sqrt{(2\pi)}}\right) q(\nu \tau)[1 + q(\nu \tau)]^{1/2} \exp[\bar{x}^2(q(\nu \tau) - 1)/2\sigma^2] \qquad (47)$$

where

$$q(\nu \tau) = 4/(4\, e^{-\nu \tau} + e^{\nu \tau})$$

A numerical investigation shows that this distribution is virtually indistinguishable from a normal distribution with mean value given by

$$\bar{\tau} = \ln(2)/\nu$$

and a coefficient of variation equal to twice that of ϕ, i.e.

$$c(\tau) = 2\sigma/\bar{x}$$

These features of the distribution $f(\tau)$ are to be expected since it follows

1. MATHEMATICS OF MICROBIAL AGE AND SIZE DISTRIBUTIONS

from the assumption of exponential growth that

$$\tau = \frac{1}{\nu} \ln\left(\frac{x_m}{x_0}\right) \qquad (48)$$

If the two variables x_m and x_0 are both distributed with coefficient of variation c (as they are in the above analysis) and they are assumed to be distributed independently, then their ratio will have a coefficient of variation given by $\sqrt{2}c$. Furthermore, the variance of the distribution of the logarithm of a positive-valued variate (x) with a small coefficient of variation is given by the square of the coefficient of variation of x (see, e.g. Powell 1964) thus

$$\text{var}(\tau) \approx \frac{2c^2(x)}{\nu^2}$$

and

$$c^2(\tau) \approx \frac{2c^2(x)}{(\nu\bar{\tau})^2} \approx \frac{2c^2(x)}{(\ln(2))^2}$$

$$\approx 4.2c^2(x) \qquad (49)$$

or

$$c(\tau) \approx 2c(x)$$

This result is, of course, dependent upon the growth law assumed for the cells. If instead we choose the form (linear growth)

$$V(x, \tau) = \beta$$

then different results for the distribution $f(\tau)$ are obtained. In this case

$$\tau = (x_m - x_0)/\beta \qquad (50)$$

and the variance of τ is β^{-2} times the sum of the variances of x_m and x_0 (see Section 2.1)

$$\text{var}(\tau) = \frac{5 \text{ var}(x_m)}{4\beta^2} \qquad (51)$$

Thus the coefficient of variation $f(\tau)$ will in this case (assuming again that x_m and x_0 are independently distributed) be given by

$$c^2(\tau) = \frac{\text{var}(\tau)}{\bar{\tau}^2} = \frac{5 \text{ var}(x_m)}{4\beta^2 \bar{\tau}^2}$$

But $\beta\bar{\tau} = \frac{1}{2}\bar{x}_m$ thus

$$c^2(\tau) = \frac{5 \text{ var}(x_m)}{\bar{x}_m^2} = 5c^2(x) \qquad (52)$$

In fact the difference between eqns (49) and (52) is too small to make a reliable distinction between the two hypotheses for the growth rate $V(x, t)$. Direct observation of cell growth during the cell cycle has led to conflicting results concerning the rate of growth as a function of the phase of this cycle, and it is extremely difficult to distinguish between exponential growth and linear growth with a doubling in growth rate near the end of the cell cycle. The results obtained above, however, do explain the observation that the coefficient of variation of cell sizes at division is approximately half that of the distribution of generation times, and lends support to the idea that the size of a cell more directly regulates its division than its age.

The other successful feature of the model proposed by Koch and Schaechter was the prediction of the generally observed negative correlation coefficients between parent- and daughter-cell generation times.

Using the notation introduced earlier and distinguishing between generations (x_{2m} = size of daughter cell at division) and again assuming exponential growth for mother and daughter cells we have

$$x_{2m} = x_{20} \exp(\nu\tau_2) = \tfrac{1}{2} x_{1m} \exp(\nu\tau_2)$$
$$= \tfrac{1}{2} x_{10} \exp[\nu(\tau_1 + \tau_2)]$$

or

$$(\tau_1 + \tau_2) = \nu^{-1}[\ln(x_{2m}) - \ln(\tfrac{1}{4}) - \ln(x_{1m})]$$

thus

$$\text{var}(\tau_1 + \tau_2) = 2\nu^{-2}\text{var}[\ln(x_m)]$$

while

$$\text{var}(\tau_1) = \text{var}(\tau_2) = \text{var}(\tau)$$
$$= 2\nu^{-2}\text{var}[\ln(x_m)] \quad \text{(see eqn (48))}$$

which implies that:

$$\rho^{(1)} = \frac{\text{var}(\tau_1 + \tau_2)}{2\,\text{var}(\tau)} - 1 = -\tfrac{1}{2}$$

By similar reasoning using τ_{s1} and τ_{s2} for the generation times of sister cells with a common mother we find that

$$\tau_{1s} = \nu^{-1} \ln\left(\frac{x'_m}{x_0}\right), \qquad \tau_{2s} = \nu^{-1} \ln\left(\frac{x''_m}{x_0}\right)$$

where x'_m and x''_m are the sizes of the sister cells at division and

$$(\tau_{1s} - \tau_{2s}) = \nu^{-1} \ln\left(\frac{x'_m}{x''_m}\right)$$

1. MATHEMATICS OF MICROBIAL AGE AND SIZE DISTRIBUTIONS

Thus

$$\operatorname{var}(\tau_{1s} - \tau_{2s}) = 2\nu^{-2}\operatorname{var}[\ln(x_m)]$$

and the correlation coefficient between sister-cell generation times is given by

$$\rho(s) = 1 - \frac{\operatorname{var}(\tau_{1s} - \tau_{2s})}{2\operatorname{var}(\tau)} = \tfrac{1}{2}$$

It is possible to improve these formulae by considering the case of unequal sizes of sister cells. Using the distribution $k(r)$ introduced in Section 4, Powell (1964) has extended the above results and shown that in this case

$$\rho(s) \leq -\rho^{(1)} \leq \tfrac{1}{2}$$

and

$$\rho(s) + \rho^{(1)} \leq 0$$

Available results for $\rho(s)$ are not as precise as those for $\rho^{(1)}$ which has been found to be close to $-\tfrac{1}{2}$ in many cases but microscopic results confirm a positive correlation between sister-cell generation times.

The above result is, of course, obtained only for the special case of exponential growth. If we use again the equation for linear growth (50) we have

$$x_{2m} - x_{20} = \beta\tau_2; \quad x_{1m} - x_{10} = \beta\tau_1; \quad x_{20} = \tfrac{1}{2}x_{1m}$$

and

$$\beta(\tau_2 + \tfrac{1}{2}\tau_1) = (x_{2m} - \tfrac{1}{2}x_{10})$$

But it is readily shown that

$$\operatorname{var}(\tau_2 + \tfrac{1}{2}\tau_1) = \operatorname{var}(\tau)(5 + 4\rho^{(1)})/4$$

and (see eqn (51))

$$\operatorname{var}(\tau) = 5\beta^{-2}\operatorname{var}(x_m)/4$$

while

$$\operatorname{var}(x_{2m} - \tfrac{1}{2}x_{10}) = \operatorname{var}(x_m) + \operatorname{var}(x_m)/16$$

thus we obtain

$$\rho^{(1)} = -\tfrac{2}{5}$$

and similarly

$$\rho^{(s)} = \tfrac{1}{5}$$

Again the available experimental evidence does not allow one to distinguish between the predictions of exponential and linear growth for the correlation coefficients. The general features of the correlations predicted by the model of Koch and Schaecter are, however, self-evident since long-lived parent cells (i.e. those which divide at a size which on average is greater than the mean) will divide to produce daughters larger than the mean size of newborn cells. These daughter cells will reach the sizes at which they will divide in times which are shorter than average since they had a size advantage over most cells when newly born. Long-lived mothers will, therefore, tend to produce short-lived daughter cells and vice versa. This is expressed quantitativeiy by a negative correlation coefficient. Similarly we expect sister cells starting at the same size to divide at times which are similar, i.e. a positive correlation can be expected between sister-cell generation times.

The exact values of the correlation coefficients will depend upon the rate of growth of the cell as a function of the phase of the cell cycle. In the case of linear cell growth, for example, the time at which any doubling in growth rate occurs (e.g. just prior to division) will modify the results obtained by neglecting this doubling above.

8. Summary

The specific models considered in the previous section illustrate the use of the formulae and approaches developed earlier in pursuing the implications of some simple hypotheses concerning the control of cellular division. In the three examples chosen, it is necessary to determine the mathematical consequences of the initial hypothesis concerning the control mechanism, and to compare these consequences with those parameters of the cell cycle which are accessible to experimental measurement.

Many other mechanisms for the control of cellular division have been proposed. A notable example is the model of Cooper and Helmstetter (1968) in which cellular division is determined by the time of completion of DNA replication. A survey of the many models which have been proposed is, however, outside the scope of this chapter. The interested reader is referred to the review by Sargent (1978) of the various hypotheses which have been advanced for the control of the cell surface area and its relationship to the cell cycle.

The mathematical techniques necessary to follow a particular hypothesis to its predictions for observable cell-cycle parameters vary greatly in complexity. In many cases, however, the techniques described or illustrated n this chapter are sufficient to yield useful results concerning the distri-

1. MATHEMATICS OF MICROBIAL AGE AND SIZE DISTRIBUTIONS

bution of cellular sizes and generation times, and of the correlations between the generation times of related cells.

References

Collins, J. F. and Richmond, M. Y. (1962). *J. Gen Microbiol.* **28**, 15–33.
Cooper, S. and Helmstetter, C. E. (1968). *J. Mol. Biol.* **31**, 519–540.
Cramer, H. (1946). "Mathematical Methods of Statistics." Princeton University Press, New Jersey.
Cramer, H. (1955). "The Elements of Probability Theory." Wiley, New York.
Eakman, J. M. (1966). Ph.D. Thesis, University of Minnesota.
Eakman, J. M., Fredrickson, A. G. and Tsuchiya, H. M. (1966). *Chem. Eng. Progr. Symp. Ser.* **62**, 37–49.
Harris, T. (1959). *In* "The Kinetics of Cellular Proliferation" (F. Stohlman, ed.), pp. 368–381. Grune and Stratton, New York.
Harvey, J. D. (1972a). *J. Gen. Microbiol.* **70**, 99–107.
Harvey, J. D. (1972b). *J. Gen. Microbiol.* **70**, 109–114.
Harvey, R. J., Marr, A. G. and Painter, P. R. (1967). *J. Bacteriol.* **93**, 605–617.
Kendall, D. G. (1948). *Biometrika* **35**, 316–330.
Koch, A. L. and Schaechter, M. (1962). *J. Gen. Microbiol.* **29**, 435–454.
Kubitschek, H. E. (1962). *Exp. Cell. Res.* **26**, 439–450.
Kubitschek, H. E. (1966). *Exp. Cell. Res.* **43**, 30–38.
Kubitschek, H. E. (1969). *In* "Methods in Microbiology" (J. R. Norris and D. W. Ribbons, eds.), Vol. 1, pp. 593–610. Academic Press, New York.
Marr, A. G., Painter, P. R. and Nilson, E. H. (1969). *Symp. Soc. Gen. Microbiol.* **19**, 237–261.
Mood, A. M. (1950). "Introduction to the Theory of Statistics." McGraw-Hill, New York.
Painter, P. R. (1975). *J. Gen. Microbiol.* **89**, 217–220.
Painter, P. R. and Marr, A. G. (1968). *Ann. Rev. Microbiol.* **22**, 519–548.
Plank, L. D. (1978). Ph.D. Thesis, University of Waikato, New Zealand.
Plank, L. D. and Harvey, J. D. (1979). *J. Gen. Microbiol.* **115**, 69–77.
Powell, E. O. (1958). *J. Gen. Microbiol.* **18**, 382–417.
Powell, E. O. (1964). *J. Gen. Microbiol.* **37**, 231–249.
Powell, E. O. (1969). *J. Gen. Microbiol.* **58**, 141–144.
Powell, E. O. and Errington, F. P. (1963). *J. Gen. Microbiol.* **31**, 315–327.
Rahn, O. (1932). *J. Gen. Physiol.* **15**, 257–277.
Sargent, S. G. (1978). *Adv. Microbiol. Physiol.* **18**, 105–176.
Sinclair, W. K. and Ross, D. W. (1969). *Biophys. J.* **9**, 1056–1070.
Trucco, E. (1970). *Bull. Math. Biophys.* **32**, 459–473.

Chapter 2

Mathematical Models in Microbiology: Mathematical Tool-kit

J. A. Howell

Department of Chemical Engineering, University College of Swansea, Swansea, U.K.

This chapter introduces some of the basic mathematical methods used in subsequent chapters of the book. The author concentrates on models of microbial population dynamics in continuous culture systems and describes how such models are derived and analysed. It is not his intention to detail all the techniques involved as such an exercise would be too lengthy for this volume. Rather, different methods are described in the context of examples in which they are used. For more detailed technical information the reader is referred to an introductory text on engineering mathematics such as that by Jeffreys (1969)—*Editorial note.*

1. Introduction

In developing mathematical models of microbial populations there is a definite procedure which is usually followed. The final model usually contains equations, probably differential equations, which have been derived from a mass balance on each of the important components of the system such as each species and each important substrate. Every balance is carried out on a well-defined spatial system and takes the following form:

net rate of increase of A = net rate of formation of A by microbial processes + net rate of transfer of A across system boundaries

The first term is simply the first derivative of A with respect to time multiplied by the volume of the system.

The second term is a kinetic term which is formulated from some assumed interaction between the system components. This assumption is often that microbial growth can be represented as a simple chemical reaction with known stoichiometry. Alternatively, a simple kinetic expression may be chosen empirically so that its behaviour is in accord with experimental observation.

The third term represents various transport and convective effects which occur in the system. It may simply represent the flows in and out of a stirred fermenter, or perhaps diffusional terms if the microbial population exists in a thick film or if substrates must first cross a phase boundary to reach the organisms, such as with hydrocarbon fermentations.

2. Kinetic Models

2.1. Exponential Growth

There are a large number of kinetic models which have been developed to describe the behaviour of populations of organisms. A large proportion of these models does not include any factor which would account for the spatial distribution of the organisms but is concerned only with the growth and decline of the population as a whole as though it were all exposed to uniform environmental conditions. Among such models are those which describe the behaviour of micro-organisms in well-mixed laboratory fermenters.

We start by formally presenting the well-known equation for exponential growth. The simplest equation for microbial growth rate applies to cells which grow at a constant rate per unit mass of cells. We call this constant the maximum specific growth rate, μ_m. If the cells are at a mass concentration X in a well-mixed batch fermenter of volume V the overall growth rate (R_x) per unit volume is $\mu_m X$ and the rate of increase of cell mass in the fermenter is $\mu_m XV$. In a small interval of time (Δt) the net increase in mass will be $\mu_m XV t$ and after this time the total number of cells will be

$$VX(t + \Delta t) = VX(t) + \mu_m VX \, \Delta t \qquad (1)$$

If Δt is very small $(X(t + \Delta t) - X(t))/\Delta t$ is the derivative dX/dt (this is in fact the *definition* of the derivative), so

$$\frac{dX}{dt} = \mu_m X \qquad (2)$$

Only in a closed batch fermenter (with no inflow or outflow of nutrients, liquids or cells) will the rate of increase of cells (dX/dt) be equal to the growth rate of (R_x). Therefore it is important that the simple equality in eqn (2) does not lead us to represent the growth rate as dX/dt. The growth rate is represented by R_x; the kinetic model we have used for growth rate is $R_x = \mu_m X$ and the rate of increase of cells in the fermenter is dX/dt. In a turbidostat the growth rate is exactly balanced by the wastage rate of

2. MATHEMATICAL MODELS IN MICROBIOLOGY

cells from the fermenter. For constant volume a mass balance on the cells gives

$$V \frac{dX}{dt} = R_x V - QX \quad (3)$$

If X is kept constant $dX/dt = 0$ so

$$R_x = \frac{Q}{V} X \quad (4)$$

which is clearly not equal to dX/dt.

Returning to the batch fermenter, if we know the inoculum size, X_0, then

$$\frac{dX}{dt} = \mu_m X \quad (5)$$

with the initial condition

$$X = X_0 \quad \text{at } t = 0 \quad (6)$$

This is easily solved as follows:

$$\frac{dX}{X} = \mu_m \, dt \quad (7)$$

$$\ln X = \mu_m t + c \quad (8)$$

where $\ln X$ indicates the natural logarithm of X. Substituting for the initial condition we obtain

$$c = \ln X_0 \quad (9)$$

thus

$$\ln \frac{X}{X_0} = \mu_m t \quad \text{or} \quad X = X_0 \exp(\mu_m t) \quad (10)$$

The simple derivation above is well known to microbiologists but deserves attention because the principles involved apply when more complex models are used. The *doubling time*, τ, in the above system can be found easily from equation (10) by letting $X = 2X_0$ and $t = \tau$:

$$\tau = \frac{\ln 2}{\mu_m} \quad (11)$$

The *time constant* or *relaxation time* of the system is not the doubling time but simply the characteristic constant of the system $(1/\mu_m)$.

2.2. Logistic Growth

If we wish to account for a limiting growth rate as population size increases we could use the logistic equation where the specific growth rate, $r_x (= R_x/X)$, depends explicitly on cell concentration:

$$r_x = \mu_m(1 - kX) \tag{12}$$

and in a batch fermenter with an inoculum X_0 the governing equations are

$$\frac{dX}{dt} = \mu_m X(1 - kX) \tag{13}$$

$$X = X_0 \quad \text{at } t = 0 \tag{14}$$

Since this equation contains a term in X^2 it is termed a *first-order non-linear differential equation* and many equations of this class cannot be integrated analytically but only numerically, usually with the aid of a computer. This particular equation can be solved by separation of variables:

$$\frac{dX}{X(1 - kX)} = \mu_m \, dt \tag{15}$$

Expanding the left-hand side by partial fractions and integrating we obtain the solution

$$\ln \frac{X(1 - kX_0)}{X_0(1 - kX)} = \mu_m t$$

or
$$X = \frac{X_0}{(1 - kX_0)\exp(-\mu_m t) + kX_0} \tag{16}$$

The final equation shows that as time increases X approaches the limiting maximum value of $1/k$. The time constant is still $1/\mu_m$ but the doubling time now depends on the particular value of X at which it is calculated.

2.3. Continuous-Flow Stirred Fermenter

The batch fermenter can be converted into a continuous-flow fermenter with constant inlet and outlet flows as depicted in Fig. 1. As it is well stirred, the outflow concentration is assumed to be identical with the concentration of the bulk of the fermenter. With no organisms present in the feed ($X_f = 0$) the equation for growth in this fermenter is

$$V \frac{dX}{dt} = r_x XV - QX \tag{17}$$

defining the dilution rate, D, as Q/V

$$\frac{dX}{dt} = R_x - DX \tag{18}$$

A *chemostat* is a continuous fermenter; operated at steady state and constant dilution rate $dX/dt = 0$ so that

$$r_x = D \tag{19}$$

In exponential growth this means that $D = \mu_m$ otherwise no steady state is possible, and even if one attempted to operate at $D = \mu_m$, small changes in either would immediately disturb the steady state and, depending on

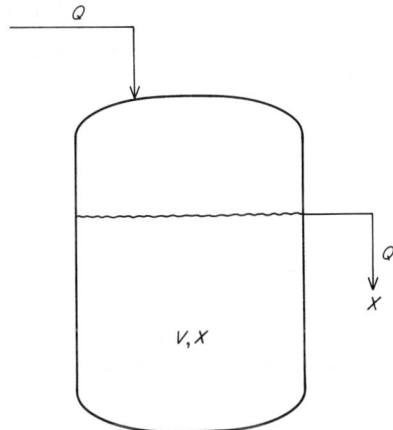

Fig. 1. Diagram of a continuous flow fermenter of volume V, flow rate Q and biomass density X.

whether $D = \mu_m$ became positive or negative, X would decline to zero (*washout*) or increase constantly. Since a constantly increasing population is not possible the exponential growth model is clearly inappropriate for a chemostat. In a turbidostat the dilution rate is controlled and thus cell mass is controlled and exponential operation is possible. In logistic growth in a chemostat

$$D = \mu_m(1 - kX) \tag{20}$$

and

$$X = \frac{1}{k}\left(1 - \frac{D}{\mu_m}\right) \qquad (21)$$

Thus $0 < D < \mu_m$ is the range of possible values for D with $0 < X < 1/k$.

2.4. Substrate Limitation

Although the logistic model has been widely used for representing microbial growth, evidence suggests that a limiting substrate can influence growth rates even at low cell concentrations. Numerous kinetic expressions exist for growth which include an explicit dependence on one or more limiting substrate concentrations. The underlying assumption of such expressions is that of all the substrates used by the cell one is present in a concentration relatively lower than the others, and growth rate depends on its concentration. If this is increased, growth rate increases until it is limited either by the concentration of another nutrient or by the innate properties of the organism.

A typical example of substrate-limited kinetics is due to Monod (1942) and takes the form

$$R_x = \frac{\mu_m S X}{K_m + S} \qquad (22)$$

where R_x is the rate of increase of the population mass concentration, S and X are substrate and biomass concentrations, respectively, and μ_m and K_m are kinetic parameters.

In such a model we shall refer to the quantities S and X as *state variables* to distinguish them from the kinetic *parameters*.

The Monod expression is not derived from any assumed stoichiometric relationship between organisms and substrates but rather occurs from an empirical approach which assumed analogous behaviour of microbial growth and Michaelis–Menten enzyme kinetics. However, a second relationship is required in any model using Monod kinetics and this relationship generally assumes that substrate is taken up in a fixed ratio to the growth rate of the biomass. The ratio of biomass growth rate to substrate uptake rate is conventionally termed the yield coefficient, but is in effect an assumed stoichiometric coefficient. In the absence of any transport terms, these stoichiometric and kinetic equations allow balances for both substrate concentration and biomass to be made. Such a situation might pertain in a well-stirred batch fermenter.

2. MATHEMATICAL MODELS IN MICROBIOLOGY

Other models can be more complex but are usually based on much the same sort of assumed dependence on a limiting substrate. Some examples are given in Table 1.

Table 1. Some microbial growth models.

Kinetics	R_x
Monod (1942)	$\dfrac{\mu_m S X}{K_m + S}$
Blackman (1905)	$kS, \ S < S^*$
	$\mu_m, \ S > S^* \left(=\dfrac{\mu_m}{k}\right)$
Monod with endogenous metabolism	$\dfrac{\mu_m S X}{K_m + S} - k_d X$
Substrate inhibited Monod Haldane	$\dfrac{\mu_m S X}{K_m + S + S^2/K_i}$
Tessier (1936)	$\dfrac{\mu_m S \exp(-S/K_i)}{K_m + S}$

2.5. Multiple Substrate Kinetics

Microbial growth depending on the concentration of more than one substrate, *multiple substrate kinetics*, is often observed. Usually a single substrate proves to be limiting and the Monod kinetic equation or a similar model is adequate to describe the behaviour of the system. Clearly, however, organisms metabolize several substrates and nutrients simultaneously. Let us take an example in which we have an organism growing on a medium containing glucose as the single carbon source, and potassium nitrate as the single nitrogen source. If carbon is limiting growth we may expect an equation of the Monod form

$$R_x = \frac{\mu_m C X}{K_c + C} \tag{23}$$

where μ_m is the maximum specific growth rate, C is the carbon concentration, K_c is the Monod coefficient for carbon-limited growth and X is the biomass concentration. If $C \gg K_c$ we expect that growth is at a maximum rate and no longer carbon limited. In this circumstance a sufficiently small nitrate concentration, N, will cause the growth rate to be less than μ_m. In other words μ_m will depend on N. It is often found that the dependence on nitrate also follows Monod kinetics and in the nitrogen-limited situation

$$R_x = \frac{\mu_m N X}{K_N + N} \tag{24}$$

where K_N is the saturation constant for nitrate. If $N = K_N$ and $C = K_c$ it is probable that both substrates will affect growth and thus a combined equation will be required that simplifies to single-substrate Monod kinetics when either substrate or nutrient is in excess. Two commonly utilized equations, each of which generalizes well to the case of many substrates, are

$$R_x = \frac{kNCX}{(K_N + N)(K_c + C)} \tag{25}$$

and

$$R_x = \frac{kX}{1 + K_N/N + K_c/C} \tag{26}$$

2.6. Structured Models

More complex models may incorporate *structure* into the biomass term. The Monod kinetic expression is unstructured because it assumes that the biomass is completely characterized by a single variable. By contrast, a structured model separates the biomass into more than one state variable. Simple structure has been incorporated into a number of models and that developed by Busby and Andrews (1975) and later used by Yeung *et al.* (1980) is one of the most straightforward, yet valuable because it demonstrates appropriate dynamic properties. Stoichiometrically it is assumed that metabolizable carbonaceous material (C) is assimilated by the cell as stored material (S) before being converted into cell mass (X) and other metabolic products. This is represented by the equation

$$C + X \rightarrow S + X \rightarrow (1 + Y)X \tag{27}$$

where Y represents a yield term. In a continuous-flow stirred fermenter these equations are combined with the general mass balance to give

$$\frac{dC}{dt} = R_c - DC \tag{28}$$

$$\frac{dS}{dt} = R_c - R_s - DS \tag{29}$$

$$\frac{dX}{dt} = YR_s - DX \tag{30}$$

where R_s is the specific rate of substrate uptake. Structured models using two biomass compartments have been developed by a number of authors including Powell (1966), Ramkrishna et al. (1967), Lee et al. (1975), Jones (1975), Busby and Andrews (1975) and Chi and Howell (1976). In each of the above cases a different biological basis was advanced for the two compartments within the single population. Nucleic acids and protein, flocculated and unflocculated, viable and active-sterile, adsorbed and viable, G-mass and D-mass were all terms used to describe the compartments. In each case the models gave a better fit to transient data than unstructured models but there has been little attempt to discriminate between them since their mathematical forms are so similar.

An example of a more complex structured model was given by Brown and Fitzpatrick (1979) in a development of Ramkrishna's earlier model (Ramkrishna et al., 1967). The model describes the growth of *Aspergillus oryzae* and structures the biomass into G-mass (nucleic acids), D-mass (proteinaceous matter) and E-mass (stored mass). The following stoichiometric relationships are assumed.

$$G + D + k_1 S \xrightarrow{k_4} 2G + D \tag{31}$$

$$G + D + k_2 S \xrightarrow{k_5} G + 2D \tag{32}$$

$$G + D + k_3 S \xrightarrow{k_6} G + D + E \tag{33}$$

$$G + D + E \xrightarrow{k_7} G + D \tag{34}$$

These equations are taken up together to give the overall stoichiometry of substrate uptake

$$R_s = (k_1 R_G + k_2 R_D + k_3 R_E) \tag{35}$$

where all the R terms represent rates of production of the mass represented by the subscript. Kinetic expressions are then needed for the rate terms on the right of the above equation. As before these are semiempirical in form and analogous to the Michaelis–Menten equation. It was assumed that there was more than one limiting substrate and each limiting substrate concentration (denoted by C_i for substrate i) appears in a Michaelis–Menten form; thus for G-mass

$$R_G = \frac{k_4 C_D C_G C_s C_o}{(k_8 + C_s)(k_9 + C_G)(k_{20} + C_o)} \tag{36}$$

where o denotes oxygen.

3. Mass Balances

3.1. Stoichiometry

For an unstructured model of a batch fermenter, the equation

$$\frac{dX}{dt} = R_x \tag{37}$$

represents the instantaneous growth rate where R_x is some function of limiting substrate concentration. The model is incomplete as it is not possible to solve the differential equation unless the population is known at some time t^*, and the substrate concentration S is known at all times.

$$\frac{dS}{dt} = -\frac{R_x}{Y} \tag{38}$$

represents a suitable equation for the rate of substrate utilization, where the parameter Y is a yield coefficient. Y, if it is a constant, represents a stoichiometric relationship between X and S. If it is not a constant a well-posed model requires that an equation be formulated expressing the dependence of Y on the state variables. The equation in S is supplemented by an initial condition specifying S at some time, preferably also t^*.

Let us suppose that $S(t^*) = S^*$ and $X(t^*) = X^*$. Adding equations (37) and (38) gives

$$\frac{d}{dt}(YS + X) = 0 \tag{39}$$

$$YS + X = YS^* + X^* \quad \text{when } t = t^* \tag{40}$$

and solving,

$$S = \frac{1}{Y}(YS^* + X^* - X) \tag{41}$$

This may now be substituted into eqn (37) using Monod kinetics (eqn (22)) to give a single first-order non-linear differential equation in X which together with the initial condition can be integrated to give the population at any subsequent time. Unfortunately, in general the complexity of such models makes integration rather hard to perform although in this case it is analytically possible to solve the equation. More complicated models are very common and present much greater difficulties.

A further use of mass balances in modelling of fermenters has been developed recently and has been used in cases where growth is assumed to be unstructured (see, e.g. Wang *et al.*, 1977a, b; Harima and Humphrey,

1980; Roels, 1980). If the elemental composition of the biomass can be assumed to be constant then a carbon balance can be performed allowing the total biomass production to be estimated from measurements of the total carbon fed and removed from the fermenter as substrate and of the total carbon dioxide evolved in the off-gas. It must be emphasized that such calculations cannot be predictive as can kinetic equations, but they can allow precise and continuous estimation of biomass, which in turn permits the accurate determination of the parameters in a kinetic expression.

3.2. Transport Terms

When it becomes necessary to add transport terms to the mass balance equations the resulting differential equations may remain as first order or, if diffusional processes are specifically included and spatial variation is accounted for, then the mass balance equations may become second-order partial differential equations with a concomitant increased difficulty of solution.

A typical situation in which transport terms are important yet the equations stay first order is a continuous, stirred fermenter, possibly a chemostat, in which a constant flow rate is maintained through the fermenter so that it contains a constant volume of liquid. The mass balance equation for a bacterial population exhibiting simple Monod kinetics, grown in a chemostat fed by nutrient but no organisms and with a dilution rate D, is

$$\frac{dX}{dt} = \frac{\mu_m S X}{K_m + S} - DX \tag{42}$$

The corresponding mass balance for the limiting substrate concentration, S, is

$$\frac{dS}{dt} = -\frac{\mu_m S X}{Y(K_m + S)} + D(S_0 - S) \tag{43}$$

where S_0 is the substrate concentration in the feed.

Much more complex systems of equations can occur when multiple species or substrates are modelled. For example, Rich (1973) discusses a model of an aquatic ecosystem in which organisms are grouped as "phytoplankton" and "consumers" and mass balances were made on detritus, organic and inorganic nitrogen and phosphorus, and organic carbons and carbon dioxide. Terms were included for the addition of all the nutrients from external sources and the phytoplankton growth rate depended on the fluctuating light intensity.

Typical equations from the system are:

$$\frac{dX_p}{dt} = \mu_p X_p - k_{rp} X_p - k_{mp} X_p - \mu_c \frac{X_c}{Y_{cp}} \quad (44)$$

$$\frac{dC_n}{dt} = k_{don} C_{on} + k_{en} X_e + k_{don} C_d + h_n - Y_{pn} \mu_p X_p \quad (45)$$

Term by term the first of the pair of equations states that the rate of increase of phytoplankton equals its growth rate less removal by endogenous respiration less its death rate less the rate of its loss due to predation by consumers. The second equation states that the rate of increase of inorganic nitrogen in the system equals the biological release from organic nitrogen plus the excretion by consumers plus the addition from external sources less the uptake by phytoplankton.

The specific growth rate of the phytoplankton is given by

$$\mu_p = \mu_{pm} \frac{I}{K_i + I} \left[\frac{C_{CO_2}}{K_{CO_2} + C_{CO_2}} \right] \left[\frac{C_p}{K_p + C_p} \right] \quad (46)$$

and

$$\mu_c = \mu_{cm} \frac{X_p}{K_{xp} + X_p} \quad (47)$$

is the consumer growth rate by predation of phytoplankton.

The exact details of the above model are not important in the present discussion but they serve to illustrate the point that a wide range of models in microbiology and ecology is mathematically very similar and can be dealt with by the same group of techniques. Certain results can then be generalized to apply to a wide group of models, which although biologically dissimilar nevertheless has a common mathematics.

3.3. Diffusion Barriers

One interesting case in which multiple substrate dependence occurs in a situation where there are transport effects is when there are large accumulations of micro-organisms in films or flocs creating a transport or diffusion barrier to the substrate. In this situation it may well occur that the surface concentration of a particular nutrient, say oxygen, is well above the value of K_{mo}, the Monod coefficient for that particular nutrient, yet its concentration can influence growth. In order to understand this apparent paradox we must visualize a cross-section through the film of organisms and draw the graph of concentration of nutrient against depth in the film.

In order for the substrates to reach the organisms they must follow a declining concentration creating a concentration gradient which drives the flux. As both nutrients are being used up we can have a situation where the concentration of one is limiting in the outer portion of the film and the concentration of another in the inner portion of the film. The inner portion may in fact be anaerobic although for nearly the whole of the film oxygen might be above the limiting concentration. This situation is quite simple to analyse and is also instructive as it shows how the governing differential equations may be derived.

Suppose that the organisms are growing according to the stoichiometric relationship between carbon, C, and oxygen, O, as follows:

$$aC + bO = \text{cells} + \text{waste} \tag{48}$$

If growth occurs at a rate R_x, then oxygen is used at a rate $b \cdot R_x$ and carbon at a rate $a \cdot R_x$.

Examining a slice within the film, oriented parallel to the film surface, of area A and thickness ΔZ, we can make a mass balance of the carbon entering and leaving the slice:

$$\text{carbon entering} - \text{carbon leaving} - \text{carbon metabolized}$$
$$= \text{carbon accumulated} \tag{49}$$

The flux of carbon into and out of the slice will be governed by Fick's Law and will appear as

$$-\mathscr{D}\frac{dC}{dz}\bigg|_z \quad \text{and} \quad -\mathscr{D}\frac{dC}{dz}\bigg|_{z+\Delta z}$$

Thus

$$\mathscr{D}A\frac{\partial C}{\partial z}\bigg|_z + \mathscr{D}A\frac{\partial C}{\partial z}\bigg|_{z+\Delta z} - aR_x A\,\Delta z = A\,\Delta z\frac{\partial C}{\partial t} \tag{50}$$

where $\mathscr{D}$ is the diffusion constant for the system in question. Dividing by $A\,\Delta z$ and letting Δz approach zero we obtain

$$\underset{\Delta z \to 0}{\text{Limit}}\; \mathscr{D}\frac{\left(\frac{\partial C}{\partial z}\big|_{z+\Delta} - \frac{\partial C}{\partial z}\big|_z\right)}{\Delta z} - aR_x = \frac{\partial C}{\partial t} \tag{51}$$

and at steady state $\partial C/\partial t = 0$ thus for a dissolved nutrient with diffusion constant $\mathscr{D}_c$,

$$\mathscr{D}_c\frac{d^2C}{dz^2} = aR_x \tag{52}$$

Similarly for oxygen

$$\mathcal{D}_o \frac{d^2 O}{dz^2} = bR_x \tag{53}$$

By definition, the first term in equation (51) is a second-order differential. At the solid support/film interface the gradients of nutrients with respect to z will be zero and at the film/liquid interface the concentration of

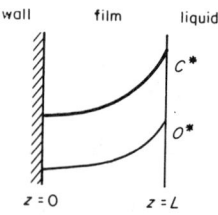

Fig. 2.

nutrients will be known, say O^* and C^*, thus with reference to Fig. 2, the above equations will have the boundary conditions

$$\frac{dC}{dz} = \frac{dO}{dz} = 0; \quad z = 0 \tag{54}$$

$$C = C^*; \quad O = O^*; \quad z = L \tag{55}$$

Then whatever kinetic expression is used for R_x it can be eliminated by combining the mass balance equations. Furthermore, if the concentrations are made dimensionless by defining $c = C/C^*$, $o = O/O^*$ and $y = z/L$ the combined equation becomes

$$\frac{C^* \mathcal{D}_c}{a} \frac{d^2 c}{dy^2} = \frac{O^* \mathcal{D}_o}{b} \frac{d_o^2}{dy^2} \tag{56}$$

$$y = 0; \quad \frac{dc}{dy} = \frac{do}{dy} = 0 \tag{57}$$

$$y = 1; \quad c = o = 1 \tag{58}$$

Thus if $k = \mathcal{D}_c C^* b / \mathcal{D}_o O^* a$ the above system can be integrated twice to give

$$k(1 - c) = 1 - o \tag{59}$$

so if $k > 1$ and if the film is deep enough then the oxygen will run out before the carbon, even if R_x has no explicit dependence on oxygen at all.

3.4. Solutions

Having obtained a set of ordinary non-linear differential equations which model the system we are studying, the set must be solved. There are a number of techniques for estimating solutions using digital computers and the user will find that most central computer installations will have a number of routines already written and stored on disc. These routines are usually appended to the users' program as subroutines and used directly. All necessary information about the model system must be supplied by the user in the main calling program and subroutines. Those potential users without training in computer programming may find complying with the above instruction difficult but there are a number of routes to success. The simplest is to find someone else to do the programming, but one must be aware that the programmer may not understand the underlying problem. Alternatively one can learn and use one of the scientific high-level languages such as FORTRAN or ALGOL or one of the very easy to use very-high-level simulation languages specifically designed for solving first-order differential equations, languages such as MIMIC, CSMP or DYNAMO. The use of the first two of these languages for microbiological modelling has been discussed by Blanch and Dunn (1974) and Nyiri (1972), respectively. These simulation languages generally integrate using a Runge–Kutta type of algorithm which is very efficient and accurate so long as the time constants associated with the problem do not differ from each other by very much. If they do, then the equations are said to be stiff and special integration routines must be used.

One example of a system of stiff equations is that describing the rate of reaction between an enzyme, E, and its substrate, S, forming an intermediate enzyme–substrate complex, ES, and yielding a product, P.

$$E + S \underset{k_{-1}}{\overset{k_1}{\rightleftharpoons}} ES \overset{k_3}{\rightarrow} E + P \tag{60}$$

This results in the familiar Michaelis–Menten expression if the first reaction is assumed to be at steady state. Initially, however, it is not at steady state and the governing differential equations are then

$$\frac{d[E]}{dt} = k_1[E][S] - (k_{-1} + k_2)[ES] \tag{61}$$

$$\frac{d[S]}{dt} = k_{-1}[ES] - k_1[E][S] \tag{62}$$

$$[E_0] = [E] + [ES] \tag{63}$$

where the square brackets represent concentration. Because the rate of the first reaction is much faster than that of the second reaction the system is stiff. The differential equations must be solved by a routine such as Gear's method or another suitable technique as discussed by Lapidus and Seinfeld (1971). The solution of the enzyme kinetic problem has been treated by Heineken et al. (1967).

Although we are not discussing enzyme reactions directly, such reactions are increasingly being introduced into microbiological models and often describe processes that occur within the cell at much faster rates than other time constants in the system. For example, the oscillations in NADH noted by Harrison et al. (1969) occurred at a period of 120 s in a fermenter with a residence time of some 7200 s, and the ratio of these two periods suggests that any applicable mathematical model might be stiff. However, for most microbiological models the procedures described by Blanch and Dunn (1974) are adequate. If they fail because of unexpected stiffness the result is that the numerical solution of the equations becomes unstable.

Although it is easy to solve such sets of equations numerically with an appropriate integration routine, unfortunately it is not in general possible to solve them analytically. In contrast to non-linear equations, linear equations can usually be solved analytically and it is thus interesting to examine the behaviour of linear equations which approximate the non-linear ones.

3.5. Estimating Model Parameters

Most of the models dealt with so far have the following features in common: they are expressed as first-order ordinary differential equations; there are stoichiometric relationships between the state variables; they are non-linear in the state variables; and at some initial point in time (usually denoted as $t = 0$) the value of the state variables (concentrations) is known. Typically when a microbial system has been described by such a model it is first used to complete itself, so to speak, and the parameters are determined by fitting the model in some way to experimental data. The model, complete with parameter values, can then be used to predict the outcome of future or hypothetical experiments. If such a prediction is successful it is generally felt that the model describes the system well enough. If the fit is unsuccessful the model must be modified until it does describe the system. This iterative process may be modified to include the simultaneous examination of a number of different models. However, such a process is time consuming and if the different models under study have such a different form that determination of the model parameters for different models requires different sets of experiments, any short cuts to the procedure become extremely welcome.

2. MATHEMATICAL MODELS IN MICROBIOLOGY

Many workers attempt first to linearize the models before fitting them to data. Thus when fitting the Monod equation it is first inverted to give a linear relationship between the inverse of the growth rate and the inverse of the limiting substrate concentration. This technique is useful in obtaining an approximate value for the maximum specific growth rate and a very approximate value for the saturation constant. Inaccuracies arise because when the equation is transformed, so are the experimental errors, and small errors at small values of the rates become very large when they are inverted. A more appropriate procedure is to use a non-linear least-squares method to fit the data, possibly using the linearized equation to give a reasonable starting point for the least-squares search technique. This method has been described by Johnson and Berthouex (1975) *inter alia*.

Successful fits are often obtained using a version of the steepest descent procedure such as Newton's or Marquardt's method (Seinfeld and Lapidus, 1974). Since in nearly all microbial models it is possible to calculate derivatives of the functions with respect to the parameters, only such methods will be discussed here. Formally the procedure may be expressed as trying to find a parameter value $k^{(1)}$ which fits the data better than a previously tried value, $k^{(0)}$. The criterion for best fit is that, Q, the sum of squares of the experimental less the predicted values for the observed variable be a minimum, i.e.

$$\text{Min } Q = \sum_{i=1}^{N} (V_{ei} - V_{pi})^2 \qquad (64)$$

where V_{ei} is the ith experimental value and V_{pi} is the ith predicted value for $i = 1, N$ values. When Q is a minimum the derivative of the sum of squares with respect to the parameter will be zero, and when it is not zero its value can be used to dictate the change in the current estimate of the value of the parameter or parameters. Formally expressed this becomes

$$k^{(1)} = k^{(0)} - \lambda \left| \frac{\partial Q}{\partial k} \right|_{k^{(0)}} \qquad (65)$$

Here λ is used to control the step size in the search routine and is adjusted on a trial-and-error basis. When nearing the minimum it will be progressively reduced. Unfortunately the above technique, the unmodified steepest descent method, although useful in one dimension has little value in higher dimensions, i.e. when there is more than one parameter to be found. In this case the solution algorithm will tend to oscillate and take a long time to converge to the minimum. In such a case the Gauss–Newton method is to be preferred. In this method the vector of derivatives $\partial S/\partial \mathbf{k}$ is premultiplied by the inverse of the matrix of second derivatives, $\mathbf{G}$, known as the Hessian matrix.

$$\mathbf{k}^{(1)} = \mathbf{k}^{(0)} - \lambda G^{-1} \frac{\partial Q}{\partial \mathbf{k}} \qquad (66)$$

where G contains components of the form $\partial^2 Q/\partial k_i \, \partial k_j$, k_i and k_j being parameters of the system. Now G can be approximated by products of the first derivatives so that

$$G_{ij} \approx 2 \frac{\partial Q}{\partial k_i} \frac{\partial Q}{\partial k_j} \qquad (67)$$

This approximation is used in several methods, including one known as Marquardt's but can suffer from the problem that the approximated matrix can cause the direction of search to be in the reverse direction from the proper one. This is countered by adding to the matrix a term γI where I is the identity matrix and γ is an arbitrary number which can be reduced to zero as the minimum is approached. All these techniques are relatively simple to program by an experienced programmer and are generally available as packages on many large computer systems. In the U.K. the programs of the Nottingham Algorithm Group (N.A.G.) (NAG Mark 7 1979) are especially valuable, and are continually being revised. If the model is expressed as differential equations the same techniques can in principle be used. In this case the trial parameters are inserted into the differential equations which are then integrated numerically. A predicted value is produced for each point in time at which there is an observed value and the sum of squares is evaluated as before. The derivatives of the sum of squares must also be expressed as differential equations and integrated with the deviations between observed and predicted values being inserted during the integration. A large set of equations must simultaneously be integrated for each trial set of values of the parameters. This requires a fairly extensive computer capability and it is tempting to try approximate methods. Since most workers now have access to appropriate computer installations they should resist such temptation and evaluate parameters by non-linear least squares. The programming effort once exerted can be used repeatedly and the results are likely to be more consistent and contain smaller errors.

A great deal of current research into models is focused on the transient behaviour of microbial systems and developing suitable dynamic models. Other chapters in this volume discuss such dynamic models and their behaviour. One purpose of this tool-kit is to show techniques which can be used to predict the behaviour of models without even solving the differential equations or evaluating parameters. In this way a certain degree of screening of models (and experiments) can take place with relatively little effort, compared with, for example, the effort required to run a 14-day continuous fermentation.

3.6. System Reduction

As an example system of equations we shall take the model of prey–predator dynamics discussed by Tsuchiya *et al.* (1972). In this model mass balances are made on the species at densities n_1 and n_2 in a well-stirred continuous fermenter.

$$\frac{dn_1}{dt} = -Dn_1 + \mu_1 n_1 - \mu_2 \frac{n_2}{Y_2} \tag{68}$$

$$\frac{dn_2}{dt} = -Dn_2 + \mu_2 n_2 \tag{69}$$

$$\frac{dS}{dt} = D(S_0 - S) - \mu_1 \frac{n_1}{Y_1} \tag{70}$$

$$\mu_1 = \frac{\mu_{1m} S}{K_1 + S}, \quad \mu_2 = \frac{\mu_{2m} n_1}{K_2 + n_1} \tag{71}$$

The species n_1 is the prey and feeds on a substrate S which is supplied to the fermenter at a concentration S_0 and a dilution rate D. The species n_1 leaves the fermenter in the effluent and is also depleted by being consumed by predator population, species n_2. Both species are assumed to follow Monod growth kinetics with respect to their respective food supplies.

Although this model is represented by three differential equations it is easy to reduce them to two by first summing all equations after scaling them by Y_1 and Y_2, the yield factors, to obtain

$$\frac{dZ}{dt} = D(S_0 - Z) \tag{72}$$

where

$$Z = S + \frac{n_1}{Y_1} + \frac{n_2}{Y_2} \tag{73}$$

This has the solution

$$Z = S_0 - (S_0 - Z_0) \exp(-Dt) \tag{74}$$

Thus any one of the two dependent variables can be expressed in terms of the other two and time. We will first examine the properties of pairs of first-order ordinary non-linear differential equations, and neglect diffusional effects.

4. Linearization of Non-linear Differential Equations

4.1. Steady States

Non-linear differential equations of the form

$$\dot{x} = \frac{dx}{dt} = f(x, y) \qquad (75)$$

$$\dot{y} = \frac{dy}{dt} = g(x, y) \qquad (76)$$

which describe the behaviour of species x and y in a chemostat, may have steady-state solutions at which dx/dt, dy/dt are zero, and thus the values of x and y do not thereafter change with time. Then the pair of equations

$$f(x, y) = 0 \qquad (77)$$

$$g(x, y) = 0 \qquad (78)$$

can be solved for the N steady states (x_i, y_i), $i = 1, \ldots, N$.

Important information regarding non-linear equations is the number and location of the steady states. For a particular set of values of the operating conditions (i.e. substrate feed concentrations, dilution rate) there may be more than one steady state and a characteristic feature of the system is the way it moves from one steady state to another. The first part of this information is contained in the behaviour of the differential equation in the immediate region of the steady states.

4.2. Taylor Series Expansion

Most functions of x at a point $x + h$ can be expressed in terms of a Taylor expansion, viz.

$$f(x + h) = f(x) + h\frac{df}{dx} + \frac{h^2}{2!}\frac{d^2f}{dx^2} + \frac{h^3}{3!}\frac{d^3f}{dx^3} + \ldots \qquad (79)$$

in which it can be shown that successive terms get smaller and smaller. Now if h is very small ($h \ll 1$) we can neglect terms containing h^2 and higher powers to obtain

$$f(x + h) \approx f(x) + h\frac{df}{dx} \qquad (80)$$

which is linearly dependent on h.

Taylor's theorem can also be applied to functions of two (or more)

2. MATHEMATICAL MODELS IN MICROBIOLOGY

independent variables thus

$$f(x + h, y + k) = f(x, y) + h\frac{\partial f}{\partial x} + k\frac{\partial f}{\partial y} + h^2\frac{\partial^2 f}{\partial x^2}$$
$$+ 2hk\frac{\partial^2 f}{\partial x \, \partial y} + k^2\frac{\partial^2 f}{\partial y^2} \tag{81}$$

Again if $h, k \ll 1$

$$f(x + h, y + k) = f(x, y) + h\frac{\partial f}{\partial x} + k\frac{\partial f}{\partial y} \tag{82}$$

which is linearly dependent on h and k.

As an example let us take the Monod–Haldane expression for substrate inhibited kinetics

$$\mu(X, S) = \frac{\mu_m SX}{K_m + S + S^2/K_i} \tag{83}$$

The system can conveniently be examined in the neighbourhood of the steady state by defining new *perturbation variables* (S', X') such that

$$S' = S - \tilde{S} \tag{84}$$

and

$$X' = X - \tilde{X} \tag{85}$$

where the tilde over a variable represents a steady-state value. The function,

$$\mu(X, S) = \mu(\tilde{X} + X', \tilde{S} + S')$$

can then be expanded in a Taylor series about $\tilde{X}, \tilde{S}$ and linearized as follows:

$$\mu(S, X) \approx \mu(\tilde{S}, \tilde{X}) + \frac{\partial \mu}{\partial S}\bigg|_{\tilde{S},\tilde{X}} S' + \frac{\partial \mu}{\partial X}\bigg|_{\tilde{S},\tilde{X}} X' \tag{86}$$

where

$$\mu(\tilde{S}, \tilde{X}) = \frac{\mu_m \tilde{X} \tilde{S}}{K_m + \tilde{S} + \tilde{S}^2/K_i}$$

$$\frac{\partial \mu}{\partial S}\bigg|_{\tilde{S},\tilde{X}} = \frac{\mu_m [K_m - \tilde{S}^2/K_i]\tilde{X}}{[K_m + \tilde{S} + \tilde{S}^2/K_i]^2}$$

and

$$\frac{\partial \mu}{\partial X}\bigg|_{\tilde{S},\tilde{X}} = \frac{\mu_m \tilde{S}}{[K_m + \tilde{S} + \tilde{S}^2/K_i]}$$

In spite of the algebraic complexity of the above expressions once the parameters and steady-state values are known the derivatives become simple numbers. For example, if $\mu_m = 0.2\,\text{h}^{-1}$, $K_m = 2\,\text{mg}\,\text{l}^{-1}$, $S = 4\,\text{mg}\,\text{l}^{-1}$, $X = 80\,\text{mg}\,\text{l}^{-1}$ and $K_i = 150\,\text{mg}\,\text{l}^{-1}$ then $\mu(S, X) = 10.48 + 0.81S' + 0.13X'$. For $S' = 1\,\text{mg}\,\text{l}^{-1}$ and $X' = 4\,\text{mg}\,\text{l}^{-1}$, the linear approximation gives $\mu = 11.82\,\text{h}^{-1}$ whereas the exact value is $11.72\,\text{h}^{-1}$.

4.3. Linearization about a Steady State

Suppose the system of equations

$$\dot{x} = f(x, y) \tag{87}$$

$$\dot{y} = g(x, y) \tag{88}$$

where $\dot{x}$ denotes dx/dt has the steady-state solution $(\bar{x}, \bar{y})$ such that

$$f(\bar{x}, \bar{y}) = g(\bar{x}, \bar{y}) = 0 \tag{89}$$

Let us now define variables x', y' such that

$$x = \bar{x} + x' \tag{90}$$

$$y = \bar{y} + y' \tag{91}$$

Substituting for these in the differential equations gives

$$\dot{x} = \dot{\bar{x}} + \dot{x}' = f(\bar{x} + x', \bar{y} + y') \tag{92}$$

and

$$\dot{y} = \dot{\bar{y}} + \dot{y}' = g(\bar{x} + x', \bar{y} + y') \tag{93}$$

Now by definition of $(\bar{x}, \bar{y})$

$$\dot{\bar{x}} = \dot{\bar{y}} = 0 \tag{94}$$

and expanding f and g by a Taylor series about $\bar{x}$ and $\bar{y}$ yields

$$f(x, y) = f(\bar{x}, \bar{y}) + \frac{\partial f}{\partial x}\bigg|_{\bar{x},\bar{y}} x' + \frac{\partial f}{\partial y}\bigg|_{\bar{x},\bar{y}} y' + O(x'^2, y'^2) \tag{95}$$

and

$$g(x, y) = g(\bar{x}, \bar{y}) + \frac{\partial g}{\partial x}\bigg|_{\bar{x},\bar{y}} x' + \frac{\partial g}{\partial y}\bigg|_{\bar{x},\bar{y}} y' + O(x'^2, y'^2) \tag{96}$$

Now if the deviation $x' \ll 1$ and the deviation $y' \ll 1$ (which means "only in the immediate region of the steady state $\bar{x}, \bar{y}$") we can neglect terms of second order in the Taylor series.

2. MATHEMATICAL MODELS IN MICROBIOLOGY

Finally using eqn (89) we obtain the linearized differential equations

$$\dot{x}' = \left.\frac{\partial f}{\partial x}\right|_{\bar{x},\bar{y}} x' + \left.\frac{\partial f}{\partial y}\right|_{\bar{x},\bar{y}} y' \tag{97}$$

$$\dot{y}' = \left.\frac{\partial g}{\partial x}\right|_{\bar{x},\bar{y}} x' + \left.\frac{\partial g}{\partial y}\right|_{\bar{x},\bar{y}} y' \tag{98}$$

These equations are linear in the deviations x', y': at $t = 0$ $x' = x_0 \ll 1$; $y' = y_0 \gg 1$.

Actually we are not limited to expansion about a steady state: we could expand the equations about any feasible pair of values (x_p, y_p). The advantage of expanding about the steady state is that if the state is stable then, after a sufficiently long time, x and y will approach the values $\bar{x}$ and $\bar{y}$. Simultaneously the values x', y' will approach zero. We can thus examine the stability property of the steady state by finding out whether x' and y' tend to zero as time tends to infinity.

4.4. Mean Value Linearization

Another type of linearization which is becoming important in a different context is mean value linearization. The advantage of a linear system of equations is such that even when the dynamics of a highly non-linear system are being studied it may make computational sense to repeatedly linearize the system about the instantaneous set of state variables. One way of doing this is mean value linearization which has been used by Howell (1980) and Stephanopoulis (1980) in their use of state observer techniques to estimate the parameters in a microbial growth model from on-line measurements of the state of the system. Mean value linearization is an iterative process in which the dynamic set of equations is converted into a linear set by partitioning each non-linear term into a product of a pseudoconstant and a single state variable. The pseudoconstant contains other state variables which are held at the previously determined value during the current iteration.

For example, if we have an equation

$$\dot{X} = kX^2 Y \tag{99}$$

$$t = 0, \quad X = X_0, \quad Y = Y_0$$

this could be converted into

$$\dot{X} = kX_0 Y_0 X \tag{100}$$

and integrated for a single time step t', the values of X and Y at that time substituted into the equation and the process repeated

$$\dot{X} = kX(t')Y(t')X \quad (101)$$

Of course this equation forms part of a system which enables $Y(t')$ to be determined also. The second equation is then repeatedly integrated over the interval $0 < t < t'$ until convergence is obtained, which generally occurs after three or four iterations. The conversion of the non-linear system to the linear system is not quite as straightforward as indicated above since for a well-behaved computation the state variables should be normalized so that their ranges and magnitudes are comparable. The success of this technique is based on the property that the mean value of a function (in this case the gradient) between two points 0 and t' can be represented by a value of the function at some point t where $0 < t < t'$.

A possible scaling might be to define $x = (X - X_0)/X_0$, $y = (Y - Y_0)/Y_0$. Then

$$\dot{x} = kY_0 X_0^2 [(1 + y(t'))(2 + x(t'))x + y + 1] \quad (102)$$

is the pseudolinear equation for x which is iteratively evaluated as part of the system. For a further discussion of this technique and its use in parameter estimation problems the reader is referred to Gill (1974).

5. Mathematical Representation

5.1. Dimensionless Equations

Linearization can be applied to the system of equations describing an organism following Monod–Haldane kinetics growing in a continuous stirred tank fermenter (CSTF). These equations are obtained by applying the general mass balance over the fermenter to both substrate (S) and biomass (X)

$$\frac{dS}{dt} = -\frac{1}{Y}\frac{\mu_m SX}{K_m + S + S^2/K_i} + D(S_0 - S) \quad (103)$$

$$\frac{dX}{dt} = \frac{\mu_m SX}{K_m + S + S^2/K_i} - DX \quad (104)$$

Before applying linearization it is instructive to make the equations dimensionless. The advantage of doing so is to produce a solution that is valid for a wide range of parameters and contains fewer parameters than the original equations. This reduces subsequent computational effort. We make the following equalities:

$$x = \frac{X}{K_m Y}, \qquad y = \frac{S}{K_m}, \qquad \tau = \mu_m t$$

$$m = \frac{D}{\mu_m}, \qquad k = \frac{K_m}{K_i}, \qquad Y_t = \frac{S_0}{K_m}$$

By substitution into eqns (103) and (104) we get

$$\frac{dy}{dt} = m(y_f - y) - \frac{xy}{1 + y + ky^2} \qquad (105)$$

$$\frac{dx}{d\tau} = -mx + \frac{xy}{1 + y + ky^2} \qquad (106)$$

It can be seen that the original six parameters have been combined into only three which completely characterize the dimensionless equations. The new parameters are two control parameters m and y_f analogous to the previous control parameters D, S_0 and one kinetic parameter, the inhibition number k. Two other kinetic parameters μ_m, K_m have been incorporated into the dimensionless state variables and the new control and kinetic parameters. The global behaviour of the system is then seen to depend only on the values of these three quantities, and a limited amount of computation will demonstrate this behaviour. As we shall see, even this amount of computation may not be needed before a general solution can be predicted.

5.2. Matrix/Vector Notation

Often it is convenient to express systems of equations in matrix/vector notation since in a complex case we might have a large number of organisms and substrates $x_1, x_2, \ldots, x_N$ and originally a set of differential equations

$$\dot{x}_1 = f_1(x_1, x_2, \ldots, x_N)$$
$$\dot{x}_2 = f_2(x_1, x_2, \ldots, x_N)$$
$$\cdot \quad \cdot$$
$$\cdot \quad \cdot \qquad (107)$$
$$\cdot \quad \cdot$$

The vector notation then sets $\mathbf{x} = (x_1, x_2, \ldots, x_N)$ and $\mathbf{f} = (f_1, f_2, f_3)$ then

$$\dot{\mathbf{x}} = \mathbf{f}(\mathbf{x}) \qquad (108)$$

represents the complete system of equations. Nothing new of a mathematical nature has been introduced here—merely a shorthand convention. The convention refers to quantities such as **x** and **f** as *vectors*.

The linearized equations are

$$\dot{x}_1 = f_{11}x_1 + f_{12}x_2 + \cdots$$
$$\dot{x}_2 = f_{21}x_1 + f_{22}x_2 + \cdots \tag{109}$$

where

$$f_{ij} = \frac{\partial f_i}{\partial x_j}$$

Eqn 109 may be represented by

$$\dot{\mathbf{x}} = \mathbf{F}\mathbf{x} \tag{110}$$

with the vector,

$$\dot{\mathbf{x}} = \left(\frac{\partial x_1}{\partial t}, \frac{\partial x_2}{\partial t}, \ldots, \frac{\partial x_N}{\partial t}\right)$$

We have introduced the further concept of a *matrix* **F** where

$$\mathbf{F} \equiv \begin{bmatrix} f_{11} & f_{12} & \cdots & f_{1N} \\ \vdots & & & \vdots \\ f_{N1} & & \cdots & f_{NN} \end{bmatrix} \tag{111}$$

A matrix is any assembly of numbers arranged into an array of two dimensions. The f_{ij} are the *components* of the matrix, f_{ij} being the component in the *i*th *row* and the *j*th *column*.

Again nothing but a shorthand has been introduced. The matrix notation has been introduced here simply for convenience in writing. There are rules for matrix manipulation which can be found in a textbook on linear algebra and these texts discuss many of the useful properties of a matrix. It is especially useful to understand the rules of matrix multiplication by a vector to obtain another vector.

In two dimensions

$$\mathbf{A}\mathbf{x} = \begin{bmatrix} a_{11} & a_{12} \\ a_{21} & a_{22} \end{bmatrix} \begin{bmatrix} x \\ y \end{bmatrix} = \begin{bmatrix} a_{11}x + a_{21}y \\ a_{21}x + a_{22}y \end{bmatrix} \tag{112}$$

2. MATHEMATICAL MODELS IN MICROBIOLOGY

In this notation the linearized pair of differential equations (97) and (98) is represented by

$$X' = Jx' \tag{113}$$

where the matrix of partial derivatives

$$J = \begin{bmatrix} \dfrac{\partial f}{\partial x} & \dfrac{\partial f}{\partial y} \\ \dfrac{\partial g}{\partial x} & \dfrac{\partial g}{\partial y} \end{bmatrix} \tag{114}$$

is generally known as the *Jacobian*.

6. Systems of Linear Differential Equations

6.1. Solution

Consider the pair of linear differential equations,

$$\dot{x} = ax + by \tag{115}$$

and

$$\dot{y} = cx + dy \tag{116}$$

From these two equations simple differentiation and repeated substitution of the first equation followed by repeated substitution of the second and first equations into the result leads to a single second-order equation,

$$\ddot{x} = a\dot{x} + b\dot{y} = a\dot{x} + bcx + bdy = a\dot{x} + bcx + d\dot{x} - adx \tag{117}$$

or

$$\ddot{x} - (a + d)\dot{x} + (ad - bc)x = 0 \tag{118}$$

The solution of the above equation is

$$x = A \exp(\lambda_1 t) + B \exp(\lambda_2 t) \tag{119}$$

where A and B are positive constants and λ_1 and λ_2 are roots (solutions) of the quadratic equation such that

$$\lambda^2 - (a + d)\lambda + (ad - bc) = 0 \tag{120}$$

This equation is more easily obtained from the determinant equation

$$\begin{vmatrix} a - \lambda & b \\ c & d - \lambda \end{vmatrix} = 0 \tag{121}$$

This equation is called the *characteristic equation* of the *matrix of coefficients*,

$$\begin{bmatrix} a & b \\ c & d \end{bmatrix}.$$

of the original differential equations. For a linearized set of differential equations the appropriate matrix would be the Jacobian of the original vector of non-linear functions evaluated at the steady state. The values of λ_1 and λ_2 which satisfy the characteristic equation are called the *eigenvalues* or characteristic values of the matrix and thus of the linear system of differential equations.

6.2. Significance of the Eigenvalues

For a system of two differential equations the eigenvalues will be the solution of a quadratic equation and thus may have both real and imaginary parts. The two eigenvalues can be expressed as a complex conjugate pair of the form,

$$\lambda = n \pm pi$$

consisting of a real part, n, and an imaginary part, pi. Here both n and p are real numbers and i is defined as $\sqrt{(-1)}$.

If, in eqn (119), the real part of the eigenvalues is negative then the terms $\exp(\lambda_1 t)$ and $\exp(\lambda_2 t)$ both decrease to zero as time increases to infinity. The value of x, therefore, tends to $A + B$, i.e. a positive constant. In this case then the system is said to be *stable* in the vicinity of the steady state. If, on the other hand, either or both of the eigenvalues has a positive real part then the solution will be *unstable* in the vicinity of the steady state. It should be emphasized that this need not imply that the system grows without limit since the linearization of the original non-linear equations was only valid in the immediate region of the steady state. The linearized system cannot give any details of the system behaviour outside this region, although, as we shall see later, some indications of the general behaviour may sometimes be obtained.

As $\exp(ix) + \exp(-ix) = \cos x$, it can be shown that any solution with imaginary parts to the eigenvalues will oscillate around the steady states. Where the real parts are negative the oscillation will be damped; where they are positive the oscillations will be undamped; and where they are zero the oscillations will be sustained. These sustained oscillations will have a behaviour which depends on the initial disturbance from the steady state.

2. MATHEMATICAL MODELS IN MICROBIOLOGY

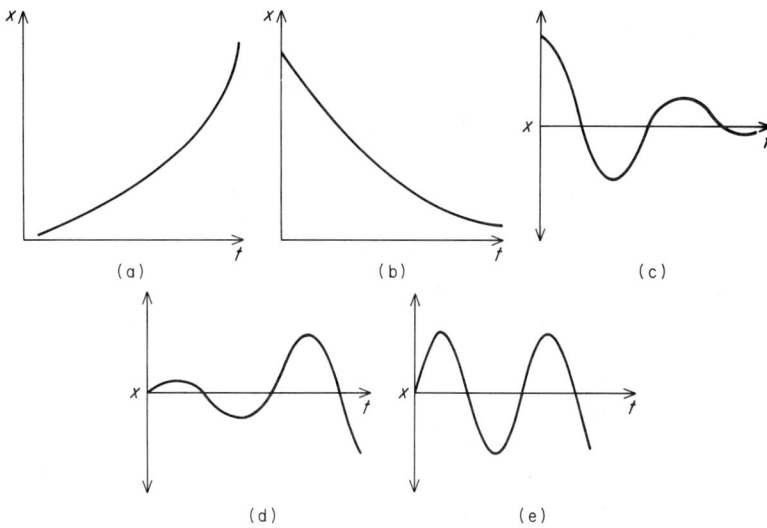

Fig. 3. Plots of variable x as a function of time showing different types of steady-state behaviour: (a) stable node; (b) unstable node; (c) stable focus; (d) unstable focus; (e) vortex point.

The time dependent behaviour of the solutions for x is shown in Fig. 3; where the sign and magnitude of the real and imaginary parts of the eigenvalues is shown in Table 2. It should be clear that as an unstable steady state may not be globally unstable as the linear system behaviour is only valid in the neighbourhood of the steady state, so a stable steady

Table 2

Eigenvalues		Behaviour	Type
Real part	Imaginary part		
Negative	$=0$	Exponential decline to 0	Stable node
Positive	$=0$	Exponential growth	Unstable node
Negative	$<\ >0$	Damped oscillation	Stable focus
Positive	$<\ >0$	Undamped oscillation	Unstable focus
Zero	$<\ >0$	Sustained oscillation	Vortex point

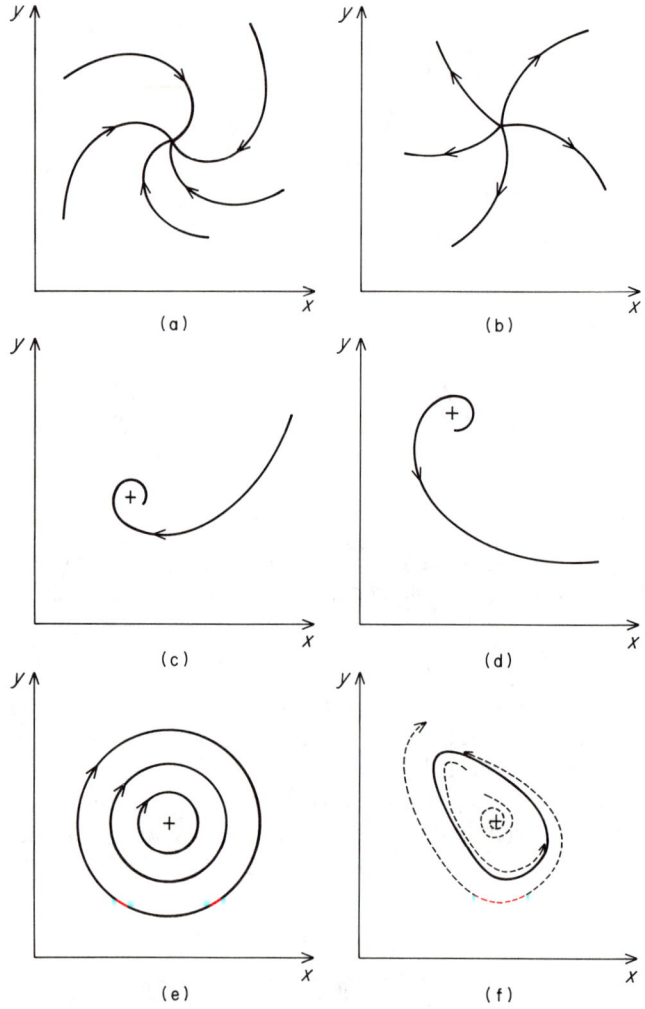

Fig. 4. Phase plane plots of x against y showing the time paths for different steady-state behaviour: (a)–(e) as in fig. 2; (f) limit cycle.

state is only shown to be stable to sufficiently small disturbances since large disturbances will not be represented by the linearized equations. Some indications of global behaviour can be obtained by making a *phase plane plot* of the integrated solutions to the non-linear equations. A phase plane plot is one in which the two dependent variables are plotted against each other for a number of time-dependent solutions (called *trajectories*) of the

2. MATHEMATICAL MODELS IN MICROBIOLOGY

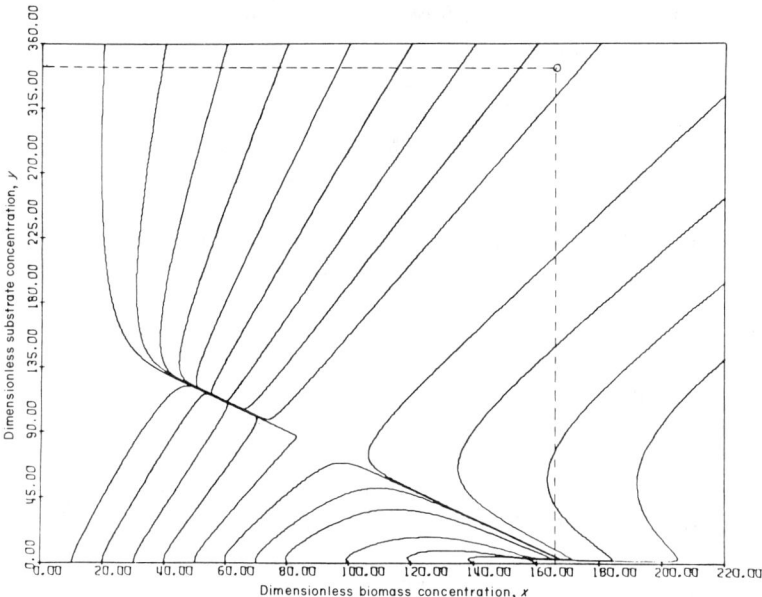

Fig. 5. Phase plane plot for a microbial growth in a chemostat exhibiting wall growth and substrate inhibition (Chi, 1974).

equations with different initial values. Figure 4 shows a number of phase plane plots which demonstrate the various types of steady-state behaviour which correspond to the conditions given in Table 2. Figure 5 shows a phase plane plot for microbial growth in a chemostat exhibiting wall growth and substrate inhibition (Chi, 1974).

6.3. Stability

Figure 4(f) displays a type of sustained oscillation which differs from that in the Lotka–Volterra equation plot of Fig. 4(e) (a vortex point). This other type of sustained oscillation is called a *limit cycle* and has a behaviour which is independent of the initial disturbance or starting point. In this it is characteristic of many microbiological systems that display oscillations; in fact it is far more common than vortex point behaviour. However, limit cycles cannot be obtained from the solution of linearized sets of differential equations but only from the original non-linear equations. Nevertheless there are theories due to Poincaré and Bendixson (Friedly, 1972) which allow the use of linearized differential equations to predict the range of parameter values where one might expect to find limit cycles.

We will examine this theory first with respect to a system of two differential equations and then show how the results may be generalized to more complex systems. Initially we shall examine the stability of the steady states in a more formal way. The conditions for stability are easily placed on a more formal basis if only two equations are involved.

The characteristic equation is obtained from the coefficient matrix

$$A = \begin{bmatrix} a & b \\ c & d \end{bmatrix} \tag{122}$$

The *trace* of the matrix is the sum of the diagonal elements, i.e.

$$\operatorname{tr} A = a + d$$

The *determinant* of the matrix A is defined only if the matrix is square and is denoted $\det A = ad - bc$. The characteristic equation (120) can then be expressed as

$$\lambda^2 - (\operatorname{tr} A)\lambda + \det A = 0 \tag{123}$$

The solution of which is

$$\lambda = \tfrac{1}{2}\{\operatorname{tr} A \pm \sqrt{[(\operatorname{tr} A)^2 - 4 \det A]}\} \tag{124}$$

If $\operatorname{tr} A > 0$, then the real part of at least one eigenvalue is greater than zero and the solution is unstable. If $\operatorname{tr} A < 0$ and $\det A < 0$ the real part of the greater eigenvalue is positive and the solution is also unstable. If $\operatorname{tr} A < 0$ and $\det A > 0$ the real part of both eigenvalues is negative and these are the necessary and sufficient conditions for stability.

Oscillations will occur in a sufficiently small region of the steady state if $\det A > 0$ and $4 \det A < (\operatorname{tr} A)$, which implies that the eigenvalues have imaginary parts, and, if a, b, c and d are real, that the eigenvalues are complex conjugates. If the steady state is unstable and there are oscillations, then it is possible that an oscillation of the limit cycle type may occur. (Vortex point sustained oscillations are not possible if the real part of the eigenvalues are non-zero.) At first sight it would appear that there is no further information to be gained from the linear equations regarding the non-linear oscillation. However, application of the Poincaré–Bendixson theory is sometimes helpful. First, if the system is globally stable (as all real systems are) and there is a unique and unstable steady state, there will be a stable limit cycle about that steady state.

If a chemostat or continuous stirred tank fermenter is described by two mass balances and thus two differential equations there can only be two eigenvalues. If the equations are suitably arranged then one of these eigenvalues will be the dilution rate and therefore the remaining eigenvalue cannot have an imaginary part since such eigenvalues must exist as complex conjugate pairs and not in isolation. This implies that such systems of equations cannot show oscillatory behaviour. Further, if oscillations, even damped ones, are observed in a single species continuous culture on a single substrate then the governing equations must contain at least two variables associated with the biomass and hence show *structure* with at least two compartments representing the biomass. As the dynamics of most cultures show some oscillatory behaviour it is thus not possible to represent them by an unstructured model. This fact is well known but not generally well understood and it is still quite common to find authors attempting to describe dynamic behaviour with unstructured models. A recent discussion on this point is given by Kossen (1979).

If we now take a two-component system showing structure and continue to analyse it by the linear techniques we are looking for situations in which the steady state is unstable and oscillatory. That is when the conditions $\det A > 0$ and $4 \det A < \operatorname{tr} A$ and $\operatorname{tr} A > 0$ hold. Now the values of the matrix A will depend on the value of the steady state which is obtained by solving the equation,

$$f(\mathbf{x}, \mathbf{u}, \mathbf{k}) = 0$$

where $\mathbf{u}$ is the vector of *control variables*, $\mathbf{x}$ is the vector of *state variables* and $\mathbf{k}$ is the vector of parameters. Control variables are those quantities under the direct control of the experimenter. State variables are those quantities, generally varying during the experiment, which describe the state of the system at any time. State variables may or may not be directly measurable but they are not under the direct control of the experimenter. Parameters are constants which represent the intrinsic properties of the system under study. It is thus possible to find situations where two of the three conditions above are satisfied and the third is an equality. Transition from an inequality to an equality as the control is varied is called *bifurcation*. We seek bifurcations since they are finite in number, and only at such bifurcations will the character of the solution change. These bifurcations can clearly occur if the number of steady states suddenly changes. Thus instead of searching in the whole of parameter and control space we need only characterize the solution at a bifurcation. It has also been shown as part of the Poincaré–Bendixson theory that for any problem a function can be defined whose sign at a bifurcation tells us whether the bifurcation branches to the right or to the left of a bifurcation point. This further reduces the labour required to characterize the solutions.

All of the above only applies in a sufficiently small neighbourhood of the steady state such that the linearization of the original equations is valid. In order to determine whether there is a reasonably sized region within which the solution is stable or unstable, a technique known as Liapunov's second method is often useful.

The theory is simply stated although it may be difficult to apply. If for any system we can define a quantity $V(x)$ which is always greater than zero (formally known as *positive definite*) except at the steady state where it is identically zero, and whose derivative with respect to time is always less than zero for some region R, then the system is *asymptotically stable* within R. For instability one may apply the same theorem to the system having first substituted $-t$ for t the time.

Although the above was simply stated it gives little information on how to find a suitable function $V(x)$. In fact there is no unique function V which satisfies the conditions. It is a problem to find a function V which satisfies the conditions for the largest region R. This problem of finding V has been discussed by many authors and is outside the scope of this chapter. The interested reader may find some useful information in Lapidus and Luus (1967) among others.

6.4. Examples

An example of the use of stability theory can be found in the works of Andrews (1968) and of Tsuchiya *et al.* (1972). In the first example Andrews used the Monod–Haldane substrate-inhibited kinetic model to describe growth of a mixed culture on phenol in a continuous stirred tank fermenter. The applicable non-linear differential equations written in dimensionless form are:

$$\dot{x} = -mx + \frac{yx}{1 + y + ky^2} \tag{125}$$

$$\dot{y} = m(y_f - y) - \frac{yx}{1 + y + ky^2} \tag{126}$$

where y represents substrate and x biomass, m is a dilution rate and k an inhibition coefficient.

There are three steady states for the above system: The trivial washout case,

$$\tilde{x}_1 = 0; \quad \tilde{y}_1 = y_f \tag{127}$$

and the solutions of

$$km\tilde{y}^2 + (m-1)\tilde{y} + m = 0 \tag{128}$$

$$\tilde{x} = y_f + \tilde{y}$$

2. MATHEMATICAL MODELS IN MICROBIOLOGY

The linearized system of differential equations about the steady state $(\bar{x}, \bar{y})$ is

$$\dot{x}' = (-m + A)x' + by' \tag{129}$$

$$\dot{y}' = -ax' - (m + b)y' \tag{130}$$

where

$$a = \frac{\bar{y}}{1 + \bar{y} + k\bar{y}^2} \quad \text{and} \quad b = \frac{1 - k\bar{y}^2}{(1 + \bar{y} + k\bar{y}^2)^2}$$

The eigenvalues can be obtained from the solution of the equation

$$\det \begin{vmatrix} -m + a - \lambda & b \\ -a & -m + b - \lambda \end{vmatrix} = 0 \tag{131}$$

adding rows and expanding

$$-(m + \lambda)(-m + b - \lambda - a) = 0 \tag{132}$$

Therefore

$$\lambda_1 = -m \tag{133}$$

$$\lambda_2 = b - a - m \tag{134}$$

Instability occurs when $\lambda_2 > 0$ or if

$$\frac{\bar{y} + \bar{y}^2 + k\bar{y}^3 - 1 + k\bar{y}^2}{(1 + \bar{y} + k\bar{y}^2)^2} > m \tag{135}$$

and using eqn (126), this implies

$$\bar{y} > \frac{1}{\sqrt{k}} \tag{136}$$

which is always so for the middle steady state. It is simple to plot the loci of the steady states on a plot of y against m^{-1}.

The critical points occur when y_f equals the larger solution of the quadratic equation (128), and below this value of $(1/m)$ the washout solution is stable; above it, it is unstable and only the low root of the quadratic is stable. Below the value of m where the quadratic has a double root, i.e. when $(1 - m)^2 < 4km^2$ only the washout solution exists. No other bifurcations arise in the system. The lowest steady state will always be stable while it exists. A similar problem was studied by Chi et al. (1974) who discussed the concept of a microbial film on the walls of the fermenter and analyzed stability in a slightly different manner. A global picture was obtained (Figs 6 and 7) of the existence of multiple steady states as a function of y_f, the substrate feed concentration, k, the inhibition parameter, and x_0, the wall growth. It was shown that in region A even for any "finite" wall growth there are multiple steady states, in region B multiplicity exists only for values of x_0 below a critical value (a_0), and in region C there are no multiple steady states.

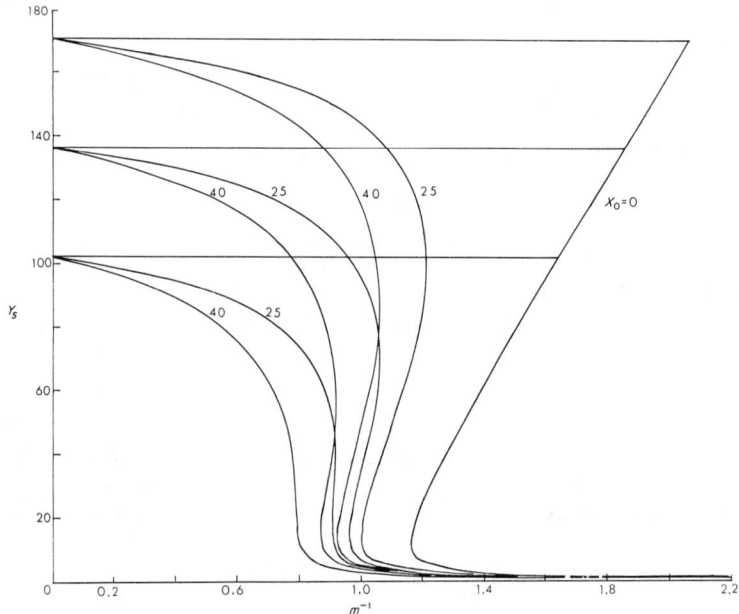

Fig. 6. Loci of steady states for substrate inhibition y versus m^{-1} and varying degrees of wall growth (x_0) and feed concentration (Chi, 1974).

A more interesting example of the use of stability theory occurs when we study the glucose–*Escherichia coli*–*Dictostelyum discoideum* predator–prey system modelled by Tsuchiya *et al.* (1972). Using eqns (68)–(71) and defining the dimensionless quantities to represent prey, $x_1 = n_1/Y_1 K_1$; predator, $z = n_2/Y_1 Y_2 K_1$; substrate, $y = S/K_1$; Monod number, $K = K_2/Y_1 K_1$; dilution rate, $m = D/u_{1m}$; relative growth $= u_{2m}/u_{1m}$ and scaled time $\tau = t_{u_{1m}}$; substitution gives

$$\dot{y} = m(y_0 - y) - \frac{yx}{1+y} \tag{137}$$

$$\dot{x} = -mx + \frac{yx}{1+y} - \frac{xz}{K+x} \tag{138}$$

$$\dot{z} = -mz + \frac{\mu xz}{K+x} \tag{139}$$

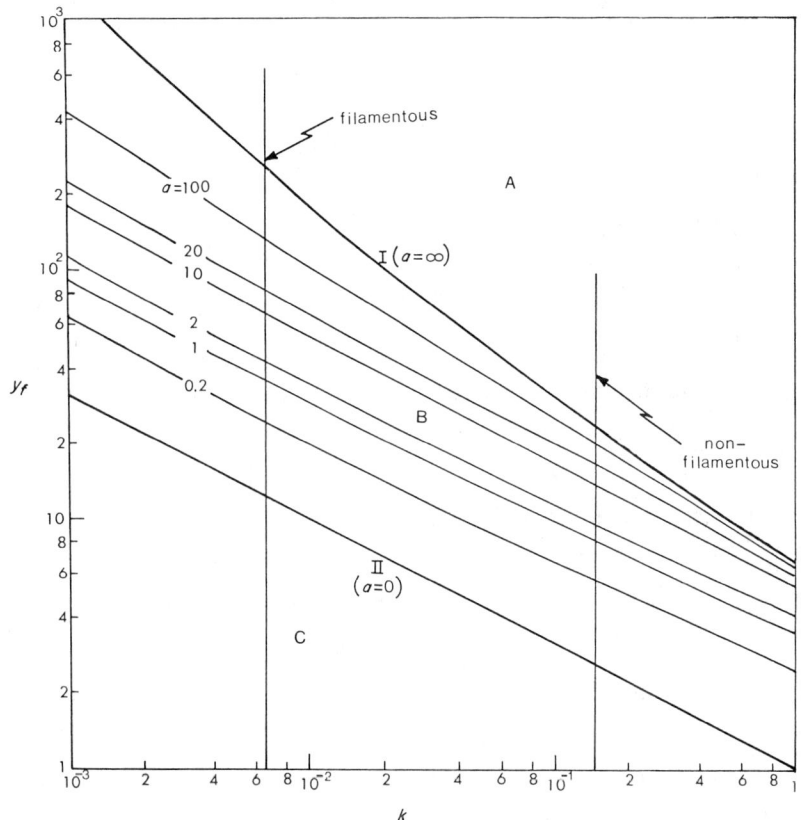

Fig. 7. Multiple steady states in a chemostat with wall growth. Region A, multiplicity always exists for any finite value of x_0; Region B, the existence of multiplicity depends on the value of x_0; Region C, no multiplicity.

There are only three steady-state solutions obtained by setting $\dot{x} = \dot{y} = \dot{z} = 0$:

(1) Total washout:

$$\tilde{x} = \tilde{z} = 0, \qquad y = y_0 \tag{140}$$

(2) Predator washout:

$$\tilde{z} = 0, \qquad \tilde{y} = \frac{m}{1-m}, \qquad \tilde{x} = y_0 - \tilde{y} \tag{141}$$

(3) Coexistence:

$$\tilde{x} = \frac{Km}{\mu - m} \tag{142}$$

$$\tilde{y} = \frac{-b + \sqrt{(b^2 + 4y_0)}}{2} \tag{143}$$

where $b = K/(\mu - m) + 1 - y_0$

$$\tilde{z} = \left(\frac{K}{\mu - m}\right)\left(m - \frac{\tilde{y}}{1 + \tilde{y}}\right) \tag{144}$$

A stability analysis can now be performed by linearizing the system of equations and forming the characteristic equation. Letting

$$a = \frac{\tilde{y}}{1 + \tilde{y}}, \quad b = \frac{\mu \tilde{x}}{K + \tilde{x}} = m, \quad c = \frac{K\mu\tilde{z}}{(K + \tilde{x})^2} \quad \text{and} \quad d = \frac{\tilde{x}}{(1 + \tilde{y})^2}$$

The characteristic equation is

$$\det \begin{vmatrix} -m + a - c - \lambda & d & -m \\ -a & -m - d - \lambda & 0 \\ c & 0 & -\lambda \end{vmatrix} = 0 \tag{145}$$

adding rows 2 and 3 to 1,

$$\det \begin{vmatrix} -m - \lambda & -m - \lambda & -m - \lambda \\ -a & -m - d - \lambda & 0 \\ c & 0 & -\lambda \end{vmatrix} = 0 \tag{146}$$

The values of λ_1 and λ_2 are the solution of

$$\lambda^2 + (m + d + c - a)\lambda + (m + d)c = 0 \tag{147}$$

Oscillations will occur when the equation has imaginary roots, that is when $(m + d + c - a)^2 < 4(m + d)c$. This focus will become unstable when $(m + d + c - a) < 0$.

Tsuchiya *et al.* determined the kinetic coefficients for their system as follows:

	μ_{im} (h^{-1})	K_i	Y_i
E. coli	0.25	5×10^{-4} mg ml^{-1}	3×10^9
D. discoideum	0.24	9×10^8 bacteria ml^{-1}	7.1×10^{-4}

2. MATHEMATICAL MODELS IN MICROBIOLOGY

For a value of $y_0 = 360$ and evaluating the three separate steady states, then:

(1) Total washout (stable) if

$$m > \frac{y_0}{1 + y_0} \quad \text{or} \quad m > 0.997$$

(2) Predator washout (stable) if

$$m > \frac{ux}{K + x} = 0.936$$

No solution if

$$m < \frac{y_0}{1 + y_0}$$

i.e. $0.936 < m < 0.997$.

(3) Coexistence (stable) with above criteria: oscillations start at $m = 0.319$ and unstable oscillations at $m = a - c - d$; $m = 0.218$.

Since all steady states are unstable the only possible solution for the non-linear system or original equations will be a *limit cycle*.

References

Andrews, J. F., (1968). *Biotech. Bioeng.* **10**.
Blackman, F. F. (1905). *Ann. Bot.* **19**, 28. (See also Dabes, J. N., Finn, R. K. and Wilke, C. R. (1973). *Biotech. Bioeng.* **15**, 1159–1177.)
Blanch, H. W. and Dunn, I. J. (1974). In "Advances in Biochemical Engineering" (T. K. Ghose, A. Fiechter and N. Blakeborough, eds.), Vol. 3. Springer-Verlag, New York, 127–166.
Brown, D. E. and Fitzpatrick, S. W. (1979). *Biotechnol. Lett.* **1**, 3–8.
Bushy, J. B. and Andrews, J. F. (1975). *J. Water Pollut. Contr. Fed.* **47**, 1055–1080.
Chi, C. T. (1974). Ph.D. Diss., State University of New York at Buffalo.
Chi, C. T. and Howell, J. A. (1976). *Biotechnol. Bioeng.* **18**, 63.
Chi, C. T., Howell, J. A. and Pawlowsky, U. (1974). *Chem. Eng. Sci.* **29**, 207–211.
Friedly, J. C. (1972). "Dynamics of Behaviour of Processes", Chap. 7. Prentice Hall, New Jersey.
Gill, P. J. (1974). Ph.D. Diss., Leeds University.
Harima, H. and Humphrey, A. E. (1980). *Biotechnol. Bioeng.* **22**, 821–831.
Harrison, D. E. F., MacLennan, D. G. and Pirt, S. J. (1969). *In* "Fermentation Advances" (D. Perlman, ed.), p. 117. Academic Press, New York.

Heineken, F., Tsuchiya, H. M. and Aris, R. (1967). *Math. Biosci.* **1**, 95–113.
Howell, A. J. (1980). Proc. VIth Int. Ferment. Symp., London, Ontario.
Jeffreys, A. (1969). "Mathematics for Engineers and Scientists." Nelson, London.
Johnson, D. B. and Berthouex, P. M. (1975). *Biotechnol. Bioeng.* **17**, 557–570.
Jones, G. L. (1975). *Progr. Water Technol.* **7**, 199–207.
Kossen, N. W. F. (1979). *In* "Microbiology: Current State, Future Prospects" (A. T. Bull, D. C. Ellwood and C. Ratledge, eds.), pp. 327–358. Cambridge University Press.
Lapidus, L. and Luus, R. (1967). "Optimal Control of Engineering Processes", Chap. 5. Blaisdell, Waltham, Massachusetts.
Lapidus, L. and Seinfeld, J. S. (1971). "Numerical Solution of Ordinary Differential Equations", Chap. 6. Academic Press, New York.
Lee, S. S., Jackman, A. L. and Schroeder, E. D. (1975). *Water Res.* **9**, 451.
Lotka, A. J. (1920). *J. Amer. Chem. Soc.* **42**, 1595.
Monod, J. (1942). "Recherches sur la Croissance des Cultures Bacteriennes" (2nd edn.). Hermann, Paris.
Nyiri, L. K. (1972). *In* "Advances in Biochemical Engineering" (T. K. Ghose, A. Fiechter and N. Blakeborough, eds.), Vol. 2. Springer-Verlag, New York.
Powell, E. O. (1966). Proc. 3rd Symp. Cont. Culture.
Ramkrishna, D., Fredrickson, A. G. and Tsuchiya, H. M. (1967). *Biotechnol. Bioeng.* **9**, 129.
Rich, L. E. (1973). "Environmental Systems Engineering". McGraw-Hill, New York.
Roels, J. A. (1980). *Biotechnol. Bioeng.* **22**, 33–53.
Seinfeld, J. S. and Lapidus, L. (1974). "Mathematical Methods in Chemical Engineering", Vol. 3, Chap. 7. Prentice-Hall, New York.
Stephanopoulis, G. (1980). Proc. VIth Int. Ferment. Symp., London, Ontario.
Svrcek, W. Y., Elliot, R. F. and Zajic, J. E. (1974). *Biotechnol. Bioeng.* **16**, 827–846.
Tiessier, C. (1936). *Ann. Physiol. Physiochim. Biol.* **12**, 527.
Tsuchiya, H. M., Fredrickson, A. G. and Aris, R. (1966). *Adv. Chem. Eng.* **6**, 125.
Tsuchiya, H. M., Drake, J. F., Jost, J. L. and Fredrickson, A. G. (1972). *J. Bacteriol.* **110**, 1147.
Wang, H. Y., Cooney, C. L. and Wang, D. I. C. (1977a). *Biotechnol. Bioeng.* **19**, 55–68.
Wang, H. Y., Cooney, C. L. and Wang, D. I. C. (1977b). *Biotechnol. Bioeng.* **19**, 69–86.
Yeung, S. Y. S., Sincic, D. and Bailey, J. E., (1980). *Water Res.* **14**, 77–83.

Chapter 3

Transients and Oscillations in Continuous Culture

A. Cunningham and R. M. Nisbet

Department of Applied Physics, University of Strathclyde, Glasgow, U.K.

Much of the early, classical work on microbial population dynamics is based on the assumption of steady state or so-called balanced exponential growth in which all the extensive properties of a culture increase at identical specific rates. Populations grow in this way only under rigorously controlled conditions. In batch cultures extensive properties such as cell number and cell mass change at different rates for much of the culture period and in the natural environment populations at steady state are rare unless they are not growing at all. The way in which microbial populations behave under non-steady-state conditions, the subject of this chapter, is therefore of considerable importance—*Editorial note.*

1. Introduction

The occurrence of transients and oscillations in microbial cultures is often regarded as a nuisance by the practising biotechnologist. Nevertheless, these phenomena are of considerable importance theoretically, since they provide information on the control of microbial system stability that cannot be obtained from steady-state observations alone. In order to maximize this information, it is necessary to observe transient behaviour under carefully controlled conditions, where the physiological state of the cells is known, or at least standardized, from one experiment to the next. Such control is possible with chemostat populations, and this chapter is consequently concerned with the modelling of transient population changes around the well-defined steady states achievable in the chemostat.

There is a considerable diversity of opinion in the literature on the complexity required in modelling chemostat populations. Authors such as Monod (1950), Droop (1968, 1978), Williams (1967) and Canale (1970) have tried to describe physiological growth states using a minimum number of parameters, and with mathematical expressions that are simple enough to be handled analytically. As a penalty for their simplicity, such models miss much of the subtlety of the growth of real microbes. On the other hand, Grenney *et al.* (1973), Sudo *et al.* (1975), Nyholm (1978) and others have proposed models that take into account the internal metabolic compartments of the microbial cell, and which consist of a large number of simultaneous differential equations. These more complex models rapidly become mathematically intractable, and rarely promote a better intuitive understanding of the systems they represent. In the present chapter, we have probably erred in the direction of oversimplifying our models; however, we feel that much of the art of successful modelling lies in choosing a model that is just sufficiently complex for the job in hand.

Since chemostat populations are large, their dynamics are essentially continuous and deterministic. The obvious way to describe such systems is by means of simultaneous-differential-equation models, and so a summary of the techniques used to investigate the stability and transient behaviour of these models is provided in Section 2. Section 3 considers the behaviour of a variety of commonly used models of growth and predation and Section 4 compares these models with the transients obtained in real microbial systems.

2. Mathematical Methods

In this section, we concentrate on deterministic models and, primarily on the grounds of learning to walk before running, neglect age-structure effects except in so far as they can be represented by single time delays in differential equations.

2.1. Differential-Equation Models

The models which we use are developed by considering the mass balance of cells and chemical matter in the chemostat, and then selecting functional forms for cell division, nutrient absorption etc. This leads to a set of differential equations of the form:

$$\dot{X}_1 = G_1(X_1, X_2, \ldots, X_m)$$
$$\dot{X}_2 = G_2(X_1, X_2, \ldots, X_m)$$
$$\vdots \qquad \vdots$$
$$\dot{X}_m = G_m(X_1, X_2, \ldots, X_m)$$

(1)

which describe the development with time of a set of m quantities, $X_1, X_2, \ldots, X_m$. Practically every computer library contains programs for

3. TRANSIENTS AND OSCILLATIONS IN CONTINUOUS CULTURE

the explicit solution of this type of equation provided we specify the values of all the parameters needed to calculate the functions G_i and the initial values of the X_i's. While such computer solutions are invaluable, much insight can be obtained and labour avoided by first performing some routine stability analysis. The procedures for this analysis have been outlined elsewhere in this volume and are extensively covered in the literature on ecological modelling (e.g. May, 1974; Nisbet and Gurney, 1982). The main steps are as follows:

(1) Identify all *feasible steady states* of the system, i.e. all steady states for which $X_i \geq 0$ for all values of i.
(2) Determine the circumstances under which the steady states are locally *stable or unstable*, i.e. determine the fate of the system after a small perturbation of the X's from their steady-state values.
(3) Where a stable steady state occurs, determine whether the return to the steady state is *oscillatory or non-oscillatory*.
(4) Where the steady state is unstable, determine whether the divergence is oscillatory or non-oscillatory. If oscillatory initiate a computer search for *limit cycles*.

Step (1) is a matter of algebraic manipulation, the ease or difficulty of which is very specific to particular models. Step (2) involves linearization about the steady state, and is most economically described using matrix notation. (The reader who finds matrices unpalatable will find the two-species case spelt out in detail in Nisbet and Gurney (1982)). We define column vectors

$$X \equiv \begin{bmatrix} X_1 \\ X_2 \\ \vdots \\ X_m \end{bmatrix} \quad \text{and} \quad G \equiv \begin{bmatrix} G_1 \\ G_2 \\ \vdots \\ G_m \end{bmatrix} \quad (2)$$

and rewrite (1) in the shorthand style

$$\dot{X} = G(X) \quad (3)$$

At a steady state (denoted by X^*) with $G(X^*) = 0$, the behaviour of the system after a small perturbation can be obtained by defining

$$x \equiv X - X^* \quad (4)$$

and expanding $G(X)$ in ascending powers of x. The result is a set of *linear* differential equations normally written in shorthand form

$$\dot{x} = Ax \quad (5)$$

where the elements of the *stability matrix* A are the partial derivatives

$$A_{ij} \equiv \left[\frac{\partial G_i}{\partial X_j}\right]_{X=X^*} \tag{6}$$

evaluated with all the X's at their steady-state values.

Local stability is established by noting (see, e.g. Sanchez 1968; Morris, 1974) that the most general solution of (5) is a sum of terms of the form

$$x = x_0 \, e^{\lambda t} \tag{7}$$

each of which must satisfy the differential equation (3). It is shown in Chapter 2 in this volume that this is only possible when λ is an *eigenvalue* of the matrix A, i.e. one of the m roots of the equation

$$\det(A - \lambda I) = 0 \tag{8}$$

These roots are in general complex numbers which we write as

$$\lambda_j = -\alpha_j \pm i\omega_j, \quad j = 1, 2, \ldots, m \tag{9}$$

and it is clear from (7) that for local stability we require all the eigenvalues of A to have negative real parts ($\alpha_j > 0$ for all j). Howell (this volume) gives conditions for this to be true.

Steps (3) and (4)—the characterization of transients and/or persisting oscillations—are our present concern, and are conveniently discussed within the mathematical framework already used in the neighbourhood stability analysis. In a locally stable system the component of eqn (7) with the smallest α ($\alpha_{\min}$, say) is the slowest to decay to zero. Over a time interval $T_R = \alpha_{\min}^{-1}$ its magnitude is reduced by a factor of almost 3. T_R is known as the *characteristic response time* and is thus a measure of the time scale over which the transient following a single perturbation remains observable. In some microbiological situations (e.g. a chemostat near the washout point) T_R can become exceptionally large, making the experimental observation of steady states extremely difficult.

If T_R is small enough for the entire transient to be observed, the main question of interest is usually whether the return to equilibrium is oscillatory or non-oscillatory. In a two-species system these cases are relatively easily distinguished by the form of the two eigenvalues of A. If both eigenvalues are real and negative, the approach of either of the x's to its steady state value is non-oscillatory (with at most one "overshoot") and we say the system is *overdamped*. If the eigenvalues are complex then from (7) and (9), the transient involves two components of the form

$$x = x_0 \, e^{-\alpha t} \, e^{\pm i\omega t} \tag{10}$$

3. TRANSIENTS AND OSCILLATIONS IN CONTINUOUS CULTURE

and the steady state is approached via a sequence of damped oscillations. The system is then said to be *underdamped*.

The condition that a two-species system is underdamped if the eigenvalues of its stability matrix are complex is mathematically sound but biologically inadequate, since oscillations which disappear very rapidly are likely to be concealed by uncertainties in experimental measurements. To quantify this, Nisbet and Gurney (1982) use the coherence number n_c which has several, essentially equivalent definitions. The most useful in the present context is

$$n_c = \frac{T_R}{P} \qquad (11)$$

the ratio of the characteristic response time to the period of the damped oscillations. The boundary between "significant" and "insignificant" oscillation is obviously arbitrary and related to the motive for seeking or avoiding oscillations in a particular system, but a rough and ready criterion is that significant oscillations are only observed if

$$n_c \geqslant 1 \qquad (12)$$

With three or more variables the distinction between overdamped and underdamped systems is more awkward to quantify. It is tempting to argue that if at least one pair of eigenvalues of the stability matrix is complex, then the return to equilibrium will contain an oscillatory component. However, this component may make no observable contribution to the transient. It is possible to concoct analytic criteria for the occurrence of significant oscillations in such systems, but the algebra is usually sufficiently messy to represent a poor return on the effort invested. A judicious mixture of analysis and computation is generally more profitable.

In a locally unstable system, an initially small perturbation does not decay to zero, and a wide variety of types of dynamical behaviour can arise. It is again useful to concentrate initially on the special case of two-species systems where the range of possibilities is limited. In such a system, the behaviour on displacement from a locally unstable steady state can be of two types:

(1) *Divergent oscillations*, following which the system *may* (depending on the nature of the non-linearities) approach a *limit cycle* on which both variables execute sustained oscillation, whose amplitude, functional form and period is independent of the initial perturbation.

(2) *Exponential divergence*, following which one or both of the variables either grows without limit, or the system approaches a new steady state or limit cycle (of large amplitude).

A limit cycle in a two-species system often contains within it an unstable steady state at which the eigenvalues of the stability matrix are complex with positive real parts. However, this is not a sufficient condition for a limit cycle, and in general the existence and form of limit cycles is most conveniently established numerically. The one exception to this generalization is two-species predator–prey models where a theorem conceived by Kolmogorov, and greatly widened in scope by Brauer (1979), guarantees for a wide class of such models that divergent oscillations indicate the existence of a limit cycle, although even here its characteristics (amplitude, period etc.) must be determined numerically.

With three or more species the range of dynamical behaviour is endless, and the pitfalls for the unwary modeller plentiful. Solutions to differential equations that diverge from an unstable steady state may approach limit cycles, but in strongly non-linear systems these may have long periods and weird shapes. Indeed, persistent aperiodic solutions (normally known as "chaotic") can occur and may be relevant in modelling laboratory insect populations. However, we know of no microbiological context in which the importance of persisting aperiodic behaviour has been established and therefore do not investigate the topic further in this chapter.

2.2. The Effect of Time Delays

Simple microbiological population models of necessity adopt an approach in which cell division is "controlled" by "macroscopic" quantities such as concentrations of intra- or extracellular limiting nutrient, and the detail of the underlying biochemistry is neglected. The price to be paid for this neglect is that differential equation models of the form of (1) may no longer be plausible. In particular there are situations where the effects of much detailed chemistry may be most economically described by invoking a time-delay τ, so that changes in a quantity A_t are determined by $B_{t-\tau}$, the value of a second quantity B at an earlier time. There are many biologically slanted accounts of the theory of the resulting *delay-differential* equations (e.g. May, 1973, 1974, 1976; Cushing, 1977; MacDonald, 1978; Maynard Smith, 1974 (pp. 43–46); Nisbet and Gurney, 1982 (Chap. 2)) so we restrict our discussion to a statement of the principal results used in the present review.

In Section 3, we consider time delays in the context of single-species chemostat models, where the total biomass at time t, X_t, satisfies an equation of the form

$$\dot{X}_t = R_t - DX_t \tag{13}$$

3. TRANSIENTS AND OSCILLATIONS IN CONTINUOUS CULTURE

in which R_t represents the net rate of recruitment to the population at time t. We shall assume that R_t depends on both X_t and $X_{t-\tau}$ so that

$$R_t = R_t(X_t, X_{t-\tau}) \tag{14}$$

With such models we are normally interested in the approach to a steady state where $X_t = X_{t-\tau} = X^*$ for all t. X^* is then the solution of the algebraic equation obtained by setting $\dot{X}_t = 0$ in (13), i.e.

$$R_t(X^*, X^*) = DX^* \tag{15}$$

We study the transient that follows a small perturbation from this steady state by setting

$$x_t \equiv X_t - X^* \tag{16}$$

Linearizing about the steady state, we obtain

$$\dot{x}_t = -ax_t - bx_{t-\tau} \tag{17}$$

where

$$a \equiv -\left[\frac{\partial R_t}{\partial X_t}\right]_{X_t = X^*, X_{t-\tau} = X^*} + D \tag{18}$$

$$b \equiv -\left[\frac{\partial R}{\partial X_{t-\tau}}\right]_{X_t = X^*, X_{t-\tau} = X^*} \tag{19}$$

In the special case where $a = 0$ (which arises with delayed regulation of *specific* growth rates) the behaviour of the system is controlled by the dimensionless product $b\tau$; it is found that the system is locally stable and overdamped if

$$0 < b\tau < 0.368 \; (=e^{-1}) \tag{20}$$

and locally stable and underdamped if

$$0.368 < b\tau < 1.57 \; (=\pi/2) \tag{21}$$

As with the simple differential equation models there is a range of parameter values for which the damped oscillations are so small as to be irrelevant, and it transpires that observable oscillations are only observed if $b\tau \geq 1$. If $b\tau$ is slightly greater than $\pi/2$, the steady state is unstable with small perturbations executing divergent oscillations with a period of approximately 4τ. In most plausible models the non-linearities will ensure that these oscillations settle down to a limit cycle, also with a period close to 4τ.

In the more interesting case where a and b are both non-zero, the behaviour of the system is controlled by the two dimensionless products $a\tau$ and $b\tau$. The algebraic detail of the stability analysis is rather tedious, but the results are readily summarized graphically as shown in Fig. 1. Much of the insight from the simple case with $a = 0$ survives; if in the absence of any time delay the system would be stable, then increasing τ leads first to damped oscillations and then possibly to divergent oscillations and limit cycles.

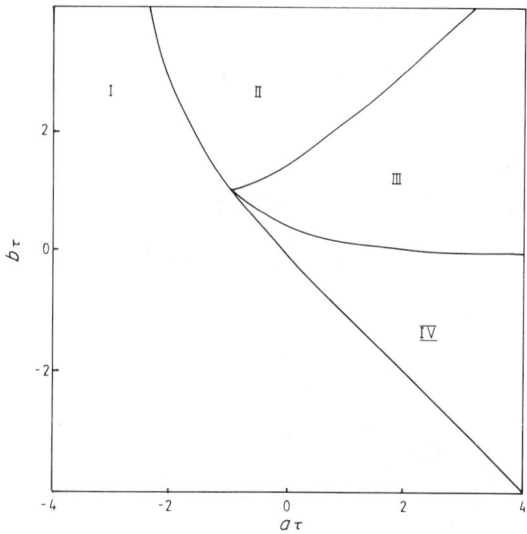

Fig. 1. Stability boundaries for eqn (217) ($\dot{x}_t = -ax_t - bx_{t-\tau}$). Regions are: I, unstable, exponentially divergent; II, unstable, divergent oscillations; III, stable, underdamped; IV, stable, overdamped.

3. The Transient Behaviour of Simple Models of Microbial Growth and Predation

The object of this section is to examine the stability and damping characteristics of some commonly cited models of growth and predation. We have selected models which are simple enough to be handled using the mathematical methods of Section 2.

3.1. Models of the Growth of a Single Species

3.1.1. *The Monod Model for Bacteria*

By considering the mass balance of cells and substrate in a chemostat, it

3. TRANSIENTS AND OSCILLATIONS IN CONTINUOUS CULTURE

is obvious that

$$\frac{dX}{dt} = (\mu - D)X \qquad (22)$$

$$\frac{dS}{dt} = D(S_r - S) - \frac{\mu X}{Y} \qquad (23)$$

where X is the mass of cells per unit volume of culture, S_r the limiting substrate concentration in the reservoir, D the fractional flow rate through the culture vessel, μ the specific growth rate of the culture and Y the yield of bacteria produced per unit mass of substrate consumed. Monod (1950) assumed that Y was constant, and that specific growth rate was related to the extracellular concentration of limiting nutrient by a hyperbola of the form:

$$\mu = \frac{\mu_{max} S}{K + S} \qquad (24)$$

where μ_{max} is the maximum value of μ, and K is the half-saturation constant for the growth process. As formulated, the Monod model describes changes in population biomass; however, it is often used to describe changes in cell number with the tacit assumption that the mean cell mass of the population is constant.

The stability of eqns (22)–(24) has been investigated by several authors, but a particularly clear account of the algebra may be found in Koga and Humphrey (1967). For parameter values yielding positive steady states, the system is locally stable and overdamped in its approach to equilibrium. Thus the Monod model is incapable of generating oscillations in response to any form of single perturbation.

One interesting point associated with the transient behaviour of this model was pointed out by Perram (1973): when the dilution rate (D) is close to its washout value the system takes a very long time to reach equilibrium. This effect is also found in other chemostat models, and it means that it is difficult to carry out transient experiments at high dilution rates. Figure 2 shows the response of the Monod model to dilution rate steps of increasing size: the Perram effect is clearly seen at high D values.

3.1.2. The Caperon–Droop Model for Phytoplankton

A considerable amount of experimental evidence suggests that the growth of algal cells is not controlled by the concentration of limiting nutrient in the environment, but rather by the concentration of that nutrient present within the cells themselves (Droop, 1968; Caperon, 1969; Fuhs, 1969;

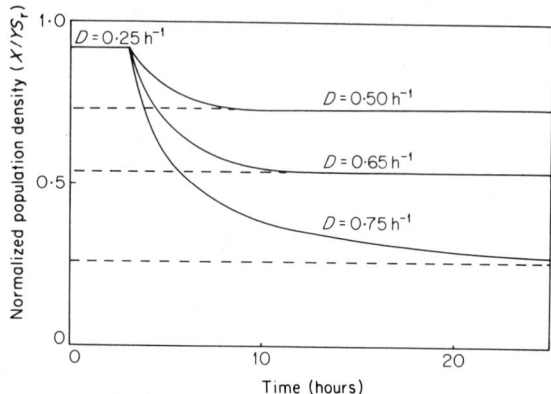

Fig. 2. Overdamped response of the Monod model (eqns (22)–(24)) to a stepped increase in dilution rate. Note the increase in the time required to each equilibrium values (indicated by broken lines) at high D values. Parameters used for these solutions: $\mu_{max} = 1\,h^{-1}$, $K_s/YS_r = 0.5$, $Y = 0.5$.

Davies, 1970; Paasche, 1973). For a population in a chemostat, this leads to three simultaneous differential equations

$$\frac{dX}{dt} = (\mu - D)X \tag{25}$$

$$\frac{dS}{dt} = D(S_r - S) - AX \tag{26}$$

$$\frac{dQ}{dt} = A - \mu Q \tag{27}$$

where A is the rate of nutrient uptake per cell, and Q the quota of limiting nutrient contained in each cell. Caperon and Meyer (1972) and Droop (1968) independently suggested that the growth function of eqn (24) should be modified as follows:

$$\mu(Q) = \begin{cases} \dfrac{\mu_{max}(Q - Q_0)}{K_Q + (Q - Q_0)}, & \text{if } Q > Q_0 \\ 0, & \text{if } Q \leq Q_p \end{cases} \tag{28}$$

where Q_0 is the minimum cell quota necessary to allow any cell division to occur, and K_Q the appropriate half-saturation constant. The transfer of nutrient from the environment to the cell interior is usually described by a second hyperbolic curve:

$$A(S) = \frac{A_{max}S}{K_S + S} \tag{29}$$

where A_{max} is the value assumed by A at infinite S. Equations (25) to (29) together comprise the Caperon–Droop model of algal growth dynamics.

At first sight, the presence of three equations in the Caperon–Droop model threatens to make stability analysis considerably more complicated. However, during normal operation of the chemostat (assuming that the system is not perturbed by the injection of extra cells of nutrient) it can be shown that:

$$S_r = S + QX \tag{30}$$

Consequently, the Caperon–Droop model can be reduced to two equations:

$$\frac{dX}{dt} = (\mu - D)X \tag{31}$$

$$\frac{dQ}{dt} = A(S) - \mu Q \quad \text{with } S = S_r - QX \tag{32}$$

Details of the linearization and stability analysis of these equations are given in Cunningham and Nisbet (1980). It transpires that the Caperon–Droop model, like the Monod model, is overdamped in its transient behaviour. Furthermore, Nisbet and Gurney (1982) have shown that this conclusion can be extended to any variation of the model in which $d\mu/dQ$ and dA/dS are positive.

Equation (28) allows the cells to store nutrient at high growth rates, which can then be used to sustain a lower growth rate in the absence of an external nutrient supply. This effect is often referred to as "luxury consumption" in the algalogical literature. One result of such storage is that the Caperon–Droop model can exhibit sharp increases in population density when the flow of medium to the reaction chamber is temporarily stopped: this effect is illustrated in Fig. 3.

3.1.3. Time Lags in Single-Species Models

In autotrophic nutrient utilization, inorganic salts are absorbed from the medium, transformed into small precursor molecules and eventually incorporated into the structure of the cell. It is likely to be the size of the precursor pool that determines the rate of synthesis of new cellular material, and the amount of new material synthesized that governs the rate of cell division. The effect of such a chain of determining factors can be crudely approximated by incorporating a discrete delay into the relationship between nutrient content and specific growth rate, so that eqn (28) is replaced by

$$\mu_t = \frac{\mu_{max}(Q_{t-\tau} - Q_0)}{K_Q + (Q_{t-\tau} - Q_0)} \tag{33}$$

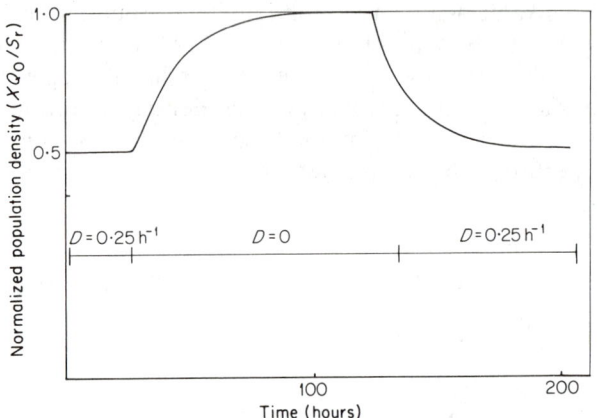

Fig. 3. The response of the Caperon–Droop model (eqns (25)–(29)) to a temporary cessation of dilution rate. Parameters used: $\mu_{max} = 0.1\,\text{h}^{-1}$, $K_Q/Q_0 = 1$, $A_{max}/Q_0 = 0.1\,\text{h}^{-1}$, $K_s/S_r = 0.01$.

where $Q_{t-\tau}$ denotes the cell quota occurring τ hours before the time t.

Methods of analysing the transient behaviour of discrete time-lag models are dealt with in Section 2.2. The application of these methods is often cumbersome in practice, and it may be necessary to simplify the model under investigation, accepting a loss of generality in the conclusions reached. For nitrogen-limited algal populations, it is found experimentally that residual nutrient levels in the chemostat are very low at all but the highest dilution rates. If we assume that these nutrient levels are effectively zero, then

$$S_r = QX \tag{34}$$

Substituting eqn (34) in (33) and linearizing about the steady state (X^*, Q^*) we find

$$\frac{dx_t}{dt} = -rx_{t-\tau} \tag{35}$$

where

$$r = \frac{S_r K_Q \mu_{max}}{X^*(K_Q + Q^* - Q_0)^2} \tag{36}$$

Thus by accepting a restriction on the range of dilution rates, we have reduced the model to a form that is already well known in the ecological literature, and whose properties have been dealt with in Section 2. The

3. TRANSIENTS AND OSCILLATIONS IN CONTINUOUS CULTURE

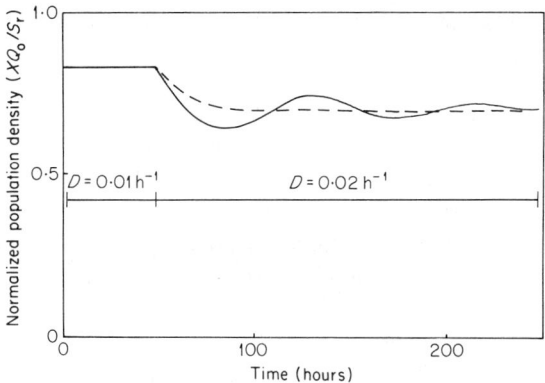

Fig. 4. Damped oscillations in the response of the Caperon–Droop model, with time lags of 0 h (---) and 20 h (——) incorporated, to a stepped increase in dilution rate. Parameters are the same as those used in Fig. 3.

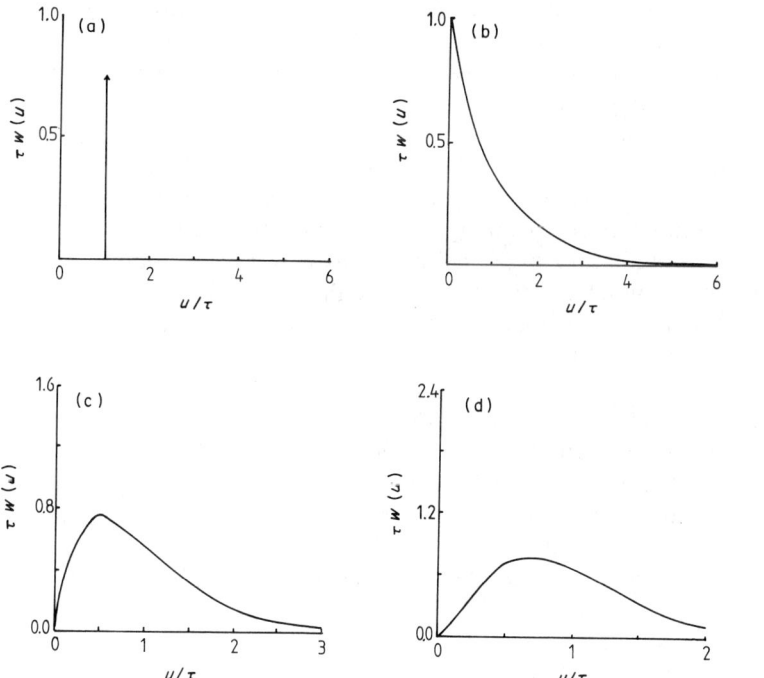

Fig. 5. Possible forms for the weighting function in eqn (37). Equations are: (a) $w(u) = \delta(u - \tau)$; (b) $w(u) = \tau^{-1}\exp(-u/\tau)$; (c) $w(u) = (4u/\tau^2)\exp(-2u/\tau)$; (d) $w(u) = (27u^2/2\tau^3)\exp(-3u/\tau)$. In each case, the mean delay is τ.

behaviour of the system is destabilized by the introduction of a time lag: observable damped oscillations are generated by $r\tau$ values greater than unity, and further increases in $r\tau$ can produce limit cycles. The extension of these conclusions to the unsimplified model can be checked numerically: Figure 4 shows the response of eqns (25)–(27), (29) and (33) to a stepped doubling in dilution rate, with time lags of 0 and 20 h incorporated.

A more sophisticated representation of the effect of nutrient history on specific growth rate may be achieved using a convolution integral instead of a simple time lag. Equation (35) would then become

$$\frac{dx}{dt} = -r \int_0^\infty w(u) x(t-u) \, du \qquad (37)$$

where $w(u)$ is a weighting function. A variety of forms have been used for $w(u)$ by Caperon (1969), MacDonald (1978) and Cunningham and Nisbet (1980); typical examples are illustrated in Fig. 5. Most of the functions that have been tested lead to transient behaviour that is qualitatively similar to that observed using the discrete time lag, though the simple exponential of Fig. 5(b) never leads to instability even at large $r\tau$ values.

Although the above analysis provides a plausible mechanism for generating oscillations, detailed studies by Cunningham and Nisbet (1980) revealed that there were serious constraints on the possible forms of the relationship between μ_t and $Q_{t-\tau}$ if such a time-delay model was to explain both non-oscillatory batch culture transients and damped oscillations in continuous culture. While it is possible that these mathematical constraints reflect some real property of the cells, it is more likely that they are a warning sign that the model is inadequate for the job in hand. The basic difficulty is probably our total neglect of age-structure when we assumed delayed regulation of the *specific* growth rate. We do not know how to resolve this problem without producing overelaborate and largely untestable models, but one possibility is to recognize that if the probability that a cell divides at time t is to be determined by its quota of limiting nutrient at time $t - \tau$, then the likeliest candidates for division at time t are those cells surviving from the total population at time $t - \tau$. We have thus studied a few models in which the *total* recruitment to the population is controlled by $X_{t-\tau}$. The local stability analysis for these models is a little more complex than before as eqn (35) is replaced by an equation of the form (17). The payoff is the removal of the pathological features (oscillations in batch culture etc.) without altering the main insight from the previous model, namely the strong destabilizing effect of the time delay.

3.2. Models of Microbial Predation

The easiest microbial food chain to analyse consists of two species in a chemostat, in which the prey is supported by a sterile nutrient solution, and the predator feeds on only one type of prey. Such food chains usually consist of a carbohydrate-limited substrate, bacterial prey and protozoan predator. By considering the mass balance of the system, it is possible to construct a general model of the system of the form:

$$\frac{dX_1}{dt} = (\mu_1 - D)X_1 - \frac{\mu_2 X_2}{Y_2} \qquad (38)$$

$$\frac{dX_2}{dt} = (\mu_2 - D)X_2 \qquad (39)$$

$$\frac{dS}{dt} = D(S_r - S) - \frac{\mu_1 X_1}{Y_1} \qquad (40)$$

where X_1 is the prey biomass density, X_2 the predator biomass density, μ_1 and μ_2 the specific growth rates of prey and predator, Y_1 the yield of prey per unit mass of substrate and Y_2 the yield of predator biomass per unit mass of prey. For the sake of simplicity eqn (39) assumes that predator death by starvation is negligible in comparison with the loss of starving cells by washout.

The stability analysis of a model consisting of three independent differential equations can be rather time consuming. Fortunately, in the present case, conservation of mass imposes the constraint that

$$S + \frac{X_1}{Y_1} + \frac{X_2}{Y_1 Y_2} = S_r \qquad (41)$$

It is therefore possible to eliminate eqn (40) from the model, and to carry out a greatly simplified analysis. The transient behaviour of eqns (38)–(40) depends critically on the functional forms assumed for the specific growth rates, μ_1 and μ_2. A range of possible forms is considered below.

3.2.1. The Lotka–Volterra Model in the Chemostat

This classic model of predation assumes that the growth of the prey is unlimited by the food supply. The specific growth rate of the predator is proportional to its rate of encounter with the prey, and continues to increase no matter how high the prey density may become. Thus, in eqns (38) and (39)

$$\mu_1 = k_1 \quad (42)$$

$$\mu_2 = k_2 X_1 \quad (43)$$

where k_1 and k_2 are constants. Equation (40) is not required in this case. These biologically unrealistic assumptions give rise to rather curious behaviour, in which the populations exhibit persistent oscillations whose amplitudes are initial-condition dependent.

3.2.2. Linear Prey and Predator Growth Functions

If we introduce specific growth-rate equations of the form

$$\mu_1 = k'_1 S \quad (44)$$

$$\mu_1 = k'_2 X_1 \quad (45)$$

then the system becomes food-limited. The result, according to Canale (1970), is to produce a model which can be under- or overdamped, but which never sustains persistent cycles.

3.2.3. Monod Growth Functions for Predator and Prey

Application of the Monod model to both stages of the food chain leads to:

$$\mu_1 = \frac{\mu_{1\max} S}{K_1 + S} \quad (46)$$

$$\mu_2 = \frac{\mu_{2\max} X_1}{K_2 + X_1} \quad (47)$$

where $\mu_{1\max}$ and $\mu_{2\max}$ are the values assumed by μ_1 and μ_2 at infinite S and X_1, respectively, and K_1 and K_2 are the appropriate half-saturation constants. These equations not only produce food limitation for both organisms; they also imply that their growth rates saturate when food becomes freely available. One obvious consequence of this saturation is a reduction in the effectiveness of the predators in controlling prey density. A full account of the stability analysis of the "Double Monod" model may be found in Canale (1970). The model can be overdamped, underdamped or limit-cycling, according to the choice of parameter values. Since this is the simplest predator–prey model that looks convincing to the experimentalist, several complete sets of parameter values have been published; and selection of these is summarised in Table 1. Figures 6–8 show typical

Table 1. Experimental parameter values for the Double Monod model.

Predator	Prey	Substrate	Y_2 (dimensionless)	$\mu_{2\max}$ (h^{-1})	K_2 (mg l^{-1})	Y_1 (dimensionless)	$\mu_{1\max}$ (h^{-1})	K_1 (mg l^{-1})	Source
Colpoda steinii	*Escherichia coli*	Glucose	0.78	0.23	6	0.45	0.9	—	Proper and Garver (1966)
Colpidium campylum	*Alcaligenes faecalis*	Asparagine	0.50	0.30	10	0.15	0.11	11	Sudo *et al.* (1975)
Tetrahymena pyriformis	*Klebsiella aerogenes*	Sucrose	0.50	0.22	12	0.45	0.6	4	Curds and Cockburn (1968)
Tetrahymena pyriformis	*Aerobacter aerogenes*	Sucrose	0.73	0.13	6	0.40	0.56	16	Canale *et al.* (1973)

Table 2. Generalized parameter set for ciliate/bacterium models.

Y_2 (dimensionless)	$\mu_{2\max}$ (h^{-1})	K_2 (mg l^{-1})	Y_1 (dimensionless)	$\mu_{1\max}$ (h^{-1})	K_1 (mg l^{-1})
0.6	0.2	9	0.4	0.5	8

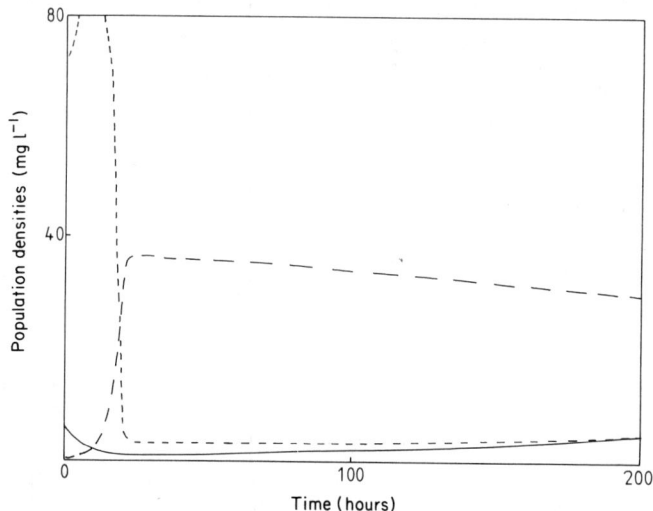

Fig. 6. Overdamped behaviour of the Double Monod model (eqns (38)–(40), (46) and (47)). $D = 0.15\,\mathrm{h}^{-1}$; $S_r = 100\,\mathrm{mg\,l}^{-1}$; other parameters are listed in Table 2; ---, substrate; ——, bacteria; ——, protozoa. Note the slow approach to the steady state caused by the Perram effect.

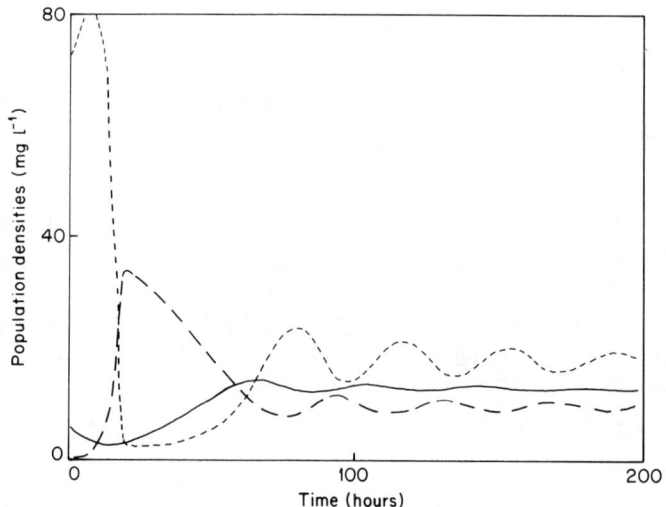

Fig. 7. Underdamped behaviour of the Double Monod model (eqns (38)–(40), (46) and (47)). $D = 0.105\,\mathrm{h}^{-1}$; other parameters as in Fig. 6; ---, substrate; ——, bacteria; ——, protozoa.

3. TRANSIENTS AND OSCILLATIONS IN CONTINUOUS CULTURE

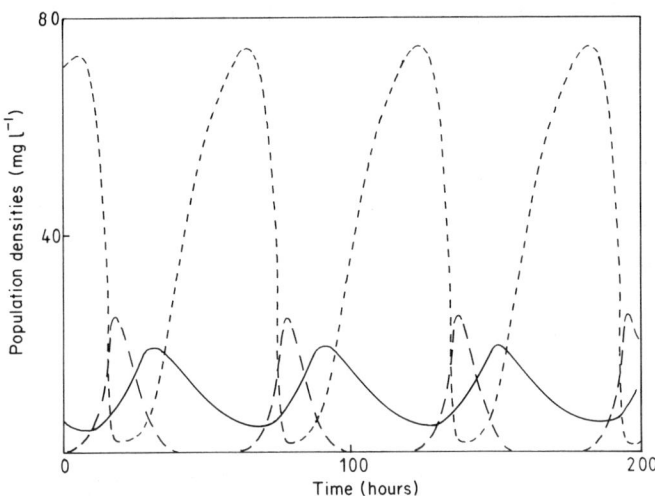

Fig. 8. Limit cycling of the Double Monod model (eqns (38)–(40), (46) and (47)). $D = 0.05 \, \text{h}^{-1}$; other parameters as in Fig. 6; (- - -, substrate; — —, bacteria; ——, protozoa).

numerical solutions of the Double Monod model, using the parameter values of Table 2, which have been chosen to lie in the midst of the experimentally determined range.

3.2.4. *The Effect of Slight Variations in the Shape of the Predator Growth Function*

A glance at the numerical solution illustrated in Fig. 8 makes it obvious that the most critical point in a predator–prey cycle occurs when the predator population density is large, and the prey population density is close to its minimum value. At this point, a small increase in predator feeding efficiency would lead to prey extinction; a small decrease in predation efficiency would curtail the downward plunge of the cycle. It is possible to explore this idea quantitatively by introducing alternative predator growth functions whose shapes differ slightly from the Monod equation at low prey densities. These shapes are illustrated in Fig. 9.

The first of these variations was suggested by Bader *et al.* (1976), namely

$$\mu_2 = \frac{\mu_{2\max} X_1^{1/2}}{K_2^{1/2} + X_1^{1/2}} \tag{48}$$

which produces a curve which initially rises more steadily than the

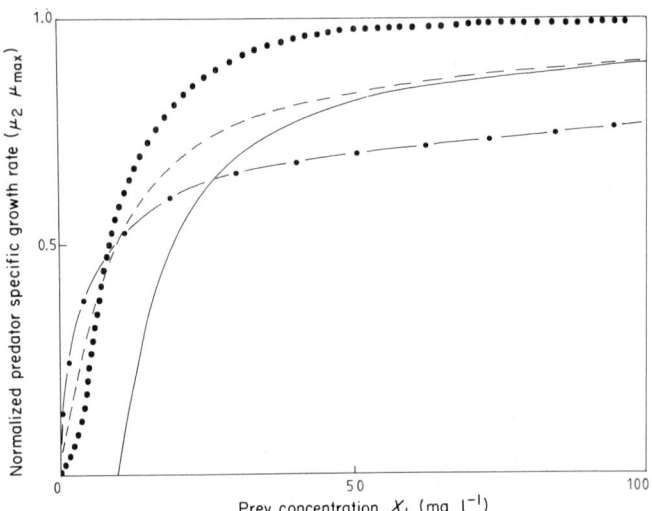

Fig. 9. Alternative functions for predator growth kinetics. The graph illustrates the Bader function (eqn (48)) (·−·−·), the Monod function (eqn (47)) (−−−), the sigmoid function (eqn (49)) (····) and the offset Monod function (eqn (50)) (———). In all cases, $K_2 = 9$ mg l^{-1}.

rectangular hyperbola of the Monod model. Bader et al. have shown that the use of such a steeply rising predator growth-rate curve greatly reduces the range of parameter values allowing predator–prey coexistence, and makes prey extinction by overexploitation the most probable outcome of the interaction.

On the other hand, the function

$$\mu_2 = \frac{\mu_{2\max} X_1^2}{K_2^2 + X_1^2} \tag{49}$$

give rise to a sigmoidal growth curve, whose key feature is a reduction in predator efficiency at low prey densities. The use of such functions produces a considerable increase in the range of parameter values which produce stable transient behaviour in the model. It is interesting to note in this context that Jost et al. (1973b) have shown that such a sigmoidal function might arise from the effects of age-structure on the feeding dynamics of ciliate populations.

Finally, it is possible to render a small proportion of the prey population inaccessible to predation by incorporating an offset in eqn (47), which leads to:

$$\mu_2 = \frac{\mu_{2\max}(X_1 - X_1')}{K_2 + (X_1 - X_1')} \tag{50}$$

3. TRANSIENTS AND OSCILLATIONS IN CONTINUOUS CULTURE 97

where X'_1 represents the prey density at which predator growth falls to zero. Again, system stability is increased.

It is obvious that the comparative effects of these growth functions could be illustrated by a series of numerical solutions of the form of Figs 6–8, but this would be a very tedious way of conveying information. The relevant points can be made more concisely by means of operating diagrams, which are discussed in the next section.

3.3 The Use of Operating Diagrams to Illustrate System Behaviour

The solution of the predator–prey models of the previous section involves the specification of about eight parameter values. Most of these parameters, however, refer to physiological characteristics of the predator and prey organisms which are not available for experimental manipulation. In the orthodox operation of the chemostat, only D and S_r comes under the control of the experimenter. It is therefore both simple and useful to construct a two-dimensional operating space for the system, and to draw contours in this space showing the boundaries between different types of transient behaviour. The position of these boundaries may be determined by referring to the stability criteria of Section 2. For example, unstable system behaviour is indicated when any eigenvalue of the characteristic matrix has a positive real part, and stable behaviour indicated when the real parts are negative. Consequently, the stable/unstable boundary occurs where the real parts are zero. In some cases, contour positions can be determined analytically; more often, however, a systematic computer search of the operating space is required.

Figures 10–12 illustrate operating diagrams for the basic model of eqns (38)–(40), using eqn (46) for the prey growth dynamics and eqns (47), (49) and (50), respectively, for the growth dynamics of the predator. As indicated in the previous section, eqns (49) and (50) both decrease the effectiveness of predation relative to the Monod model at low prey densities. The incorporation of either equation produces a considerable increase in the stable operating area of the system. An operating diagram for the same basic model, with eqn (48) incorporated, may be found in Bader et al. (1976): in this case, a large area of predator-mediated prey-extinction appears in the diagram.

Unfortunately, the operating diagrams of Figs 10–12 are difficult to interpret from the point of view of the experimental biologist. These difficulties arise from two main causes. First, the system described by eqns (38)–(40) is deterministic and continuous, whereas the behaviour of real populations at low numerical levels is discrete and often stochastic.

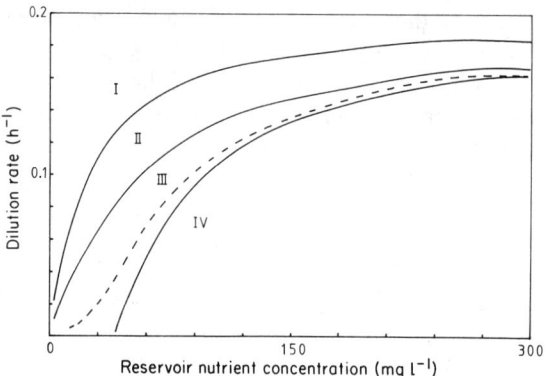

Fig. 10. Operating diagram for the Double Monod model (eqns (38)–(40), (46) and (47)) incorporating the parameter values of Table 2. Regions are: I, predator washout; II, overdamped; III, underdamped; IV, limit cycles. The broken line (– – –) is the contour for a coherence number of one.

Consequently the model may generate persistent oscillations of very large amplitude when operating at parameter values that would actually lead to the extinction of real populations. Since this effect takes place far from equilibrium, little help is offered by the linearized stability methods of Section 2. The only remedy is cautious numerical solution if one is considering a system working in the unstable region.

The second difficulty of interpretation stems from the fact that although it is mathematically possible to detect very small oscillations in model populations, oscillations observed experimentally usually have to be large

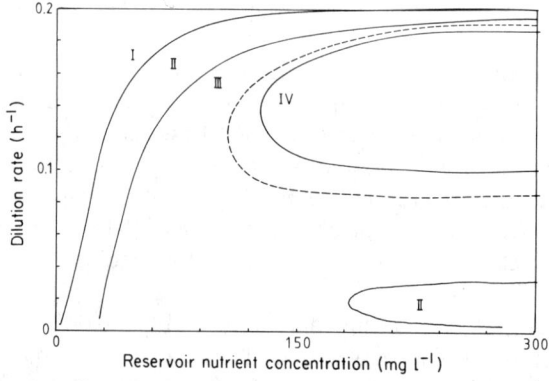

Fig. 11. Operating diagram for the basic predator–prey model (eqns (38)–(40)) with Monod prey growth kinetics (eqn (46)) and sigmoidal predator growth kinetics (eqn (49)). Contours and regions as in Fig. 10.

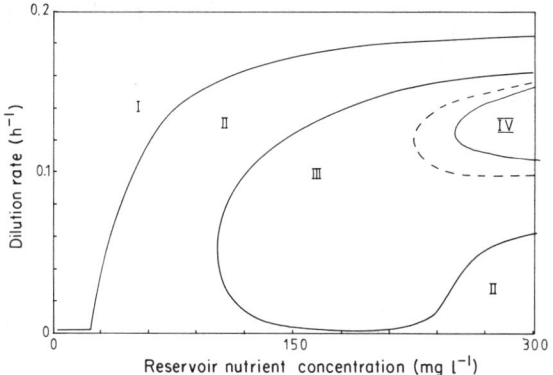

Fig. 12. Operating diagram for the basic predator–prey growth model (eqns (38)–(40)) with Monod prey growth kinetics (eqn (46)) and offset predator growth kinetics (eqn (50)). Contours and regions as in Fig. 10.

enough to appear above a residual noise level. This problem is dealt with in Section 2, where we introduced the coherence number as a measure of the size and duration of transient oscillations. Conventionally, a coherence number of one may be taken to indicate the boundary between observable and non-observable oscillations. Such a boundary has been added as a broken line in Figs 10–12. It immediately becomes obvious that most of the complicated structure of these operating diagrams is of no experimental relevance, since it occurs in areas where the distinction between over- and underdamping is purely academic. Thus the addition of coherence number contours is an extremely useful aid to the interpretation of operating diagrams.

4. Transients and Oscillations Observed Experimentally

4.1. Single Species

The Monod model for the growth of bacteria (eqns (22)–(24)) is overdamped in transient behaviour, and this overdamping persists even when internal nutrient storage is incorporated (eqn (28)). It is therefore reassuring to find that most transient experiments conducted on yeast and bacteria have failed to discover undamped behaviour.

Good transient data for algal cultures is hard to come by, but that of Williams (1971) and Cunningham and Maas (1978) shows heavily damped oscillations in cell numbers, and that of Giddings (1977) may do so. Such oscillations can only be generated by the simple models of Section 3.1 if a time lag is introduced in the response of growth rate of fluctuations in the internal nutrient pool (eqn (33)). There is little evidence for the exact form that such a lag may take, but an experimental demonstration of its existence may be found in Cunningham and Maas (1978).

4.2. Prey and Predator

Although the microbial predator–prey literature is rich in data illustrating persistent oscillations, much of this data does not come from chemostat cultures and therefore falls outside the scope of the present chapter. Predator–prey chemostat oscillations may be found, however, in the publications of Tsuchiya *et al.* (1972), van den Ende (1973), Jost *et al.* (1973a, b), Sudo *et al.* (1975), Bader *et al.* (1976) and Dent *et al.* (1976). It is a little disappointing, but very enlightening, to find that none of these experimental systems corresponds exactly to the models of Section 3.2. Each predator–prey combination seems to exhibit perculiarities of behaviour which make the formulation of a generalized predation model extremely difficult. Some of these peculiarities are listed below.

4.2.1. *Prey Flocculation*

Sudo *et al.* (1975) find that the bacterium *Alcaligenes faecalis* flocculates in the presence of high concentrations of the predatory ciliate *Colpidium campylum*. This flocculation renders the prey inaccessible to the predator.

4.2.2. *Wall Growth*

Most protozoan cultures must be stirred gently, and there is always a danger of prey cells adhering to the walls of the culture vessel. The experiments of Jost *et al.* (1973a) were terminated by wall growth, and Bonomi and Frederickson (1976) suggest that even a small amount of such growth can severely alter the dynamics of the system, by providing a reservoir of prey cells which are immune to predation.

4.2.3. *Predator Encystment and Light-Sensitive Prey*

Bader *et al.* (1976) give details of a very complicated interaction between ciliate predators (*Colpoda steinii*) and algal prey (*Anacystis nidulans*). The ciliates encyst when starving; the prey react unfavourably to sudden changes in light intensity. Since such light changes can be the consequence of a reduction in self-shading caused by predation, the overall dynamics require a great deal of experimental disentangling.

4.2.4. Metabolic Differentiation in the Predator

Amoeboid cells of the slime-mould *Dictyostelium discoideum* seem to alter their feeding strategy after about 12–14 days of continuous culture (Tsuchiya *et al.*, 1972; Dent *et al.*, 1976). This change may be linked to the very complicated life cycle of these organisms. At the least, such behaviour requires the use of a time-dependent predator growth function: Bazin and Saunders (1978) have made an interesting attempt to model the system using catastrophe theory which is discussed in detail in Chapter 4 of this volume.

4.2.5. Evolution of Prey and Predator Characteristics

There is a danger in long-term experiments that micro-organisms may be under sufficient selective pressure to change their genetic characteristics. Van den Ende (1973) suggests that this may explain the loss of mucoid capsules by the bacterium *Klebsiella erogenes* when cultured in a chemostat with *Tetrahymena pyriformis*.

4.2.6. Predator Size Fluctuations

All of the models considered in Section 3 were biomass models; similar equations have been used to describe population numbers, with the tacit assumption of constant cell size. This assumption can rarely be justified: ciliate predators change size drastically according to their nutritional state, and algal cells have a large capacity for storing energy-rich metabolites. It seems likely that microbial ecologists will be driven to measure, and eventually to model, cell size and number variations simultaneously.

None of the details listed above were foreseen in the simple predator–prey models of Section 3. However, these models have led to some ideas about the way experimental systems might function. We can consider the straightforward Double Monod model as a useful reference system, against which the behaviour of other systems can be judged: this model only allows stable predator–prey coexistence within a narrow band of parameter values (Fig. 10). Our analysis of models incorporating eqns (48)–(50) suggest that any factor that increases the specific growth rate of the predator at low prey densities, relative to the hyperbolic function of the Monod model, is liable to further reduce the stable operating area of the system. On the other hand, any factor causing a relative decrease in the predator specific growth rate at low prey densities is liable to increase the stable operating area. We have therefore gained a qualitative feel for the effects of prey self-defence mechanims such as clumping and wall growth, and also for at least some of the effects of predator age structure (Jost *et al.*, 1973b) and predator maintenance energy requirement (Luckinbill, 1973), without necessarily being able to predict these effects quantitatively.

5. Conclusions

It is very difficult to obtain an exact fit between simple models of microbial growth and the actual transient behaviour observed experimentally. However, by making a suitable choice of growth function, it is possible to obtain good approximations in a wide variety of cases. Such approximate modelling has much to offer: it helps identify the critical factors governing system stability, and may reduce the amount of systematic experimentation required to establish the system's properties. Although the types of transient behaviour exhibited by simple models are relatively few, the actual behaviour occurring at a given dilution rate and reservoir nutrient concentration is highly dependent on the function used to relate growth to food consumption. Unfortunately, the choice of this function at present is often an arbitrary one; despite the lapse of 30 years since Monod's pioneering work on bacteria, we still lack a clear description of the relationship between nutrient history and growth rate in eukaryotic cells.

References

Bader, F. G., Tsuchiya, H. M. and Frederickson, A. G. (1976). *Biotechnol. Bioeng.* **18**, 333–348.
Bazin, M. J. and Saunders, P. T. (1978). *Nature, Lond.* **275**, 52–54.
Bonomi, A. and Frederickson, A. G. (1976). *Biotechnol. Bioeng.* **18**, 239–252.
Brauer, F. (1979). *Theor. Pop. Biol.* **15**, 268–273.
Canale, R. P. (1970). *Biotechnol. Bioeng.* **12**, 353–378.
Canale, R. P., Lustig, T. D., Kehrberger, P. M. and Salo, J. E. (1973). *Biotechnol. Bioeng.* **15**, 707–728.
Caperon, J. (1969). *Ecology* **50**, 188–192.
Caperon, J. and Meyer, J. (1972). *Deep Sea Res.* **19**, 619–632.
Cunningham, A and Maas, P. (1978). *J. Gen. Microbiol.* **104**, 227–231.
Cunningham, A. and Nisbet, R. M. (1980). *J. Theor. Biol.* **84**, 189–203.
Curds, C. R. and Cockburn, A. (1968). *J. Gen. Microbiol.* **54**, 343–358.
Cushing, J. M. (1977). "Integrodifferential Equations and Delay Models in Population Dynamics". Lecture Notes in Biomathematics, Vol. 20. Springer-Verlag, Berlin, Heidelberg and New York.

3. TRANSIENTS AND OSCILLATIONS IN CONTINUOUS CULTURE

Davies, A. G. (1970). *J. Mar. Biol. Ass. U.K.* **50**, 65–86.
Dent, V. E., Bazin, M. J. and Saunders, P. T. (1976). *Arch. Microbiol.* **109**, 187–194.
Droop, M. R. (1968). *J. Mar. Biol. Ass. U.K.* **48** 689–733.
Droop, M. R. (1978). *Limnol. Oceanogr.* **23**, 283–285.
Fuhs, G. W. (1969). *J. Physiol.* **5**, 312–321.
Giddings, J. M. (1977). *Limnol. Oceanogr.* **22**, 911–918.
Grenney, W. S., Bella, D. A. and Curl, H. C., Jr. (1973). *Biotechnol. Bioeng.* **15**, 331–358.
Jost, J. L., Drake, J. F., Frederickson, A. G. and Tsuchiya, H. M. (1973a). *J. Bacteriol.* **113**, 834–840.
Jost, J. L., Drake, J. F., Tsuchiya, H. M. and Frederickson, A. G. (1973b). *J. Theor. Biol.* **41**, 461–484.
Koga, S. and Humphrey, A. E. (1967). *Biotechnol. Bioeng.* **9**, 375–386.
Luckinbill, L. S. (1973). *Ecology* **54**, 1320–1327.
MacDonald, N. (1978). "Time Lags in Biological Models". Lecture Notes in Biomathematics, Vol. 27. Springer-Verlag, Berlin, Heidelberg and New York.
May, R. M. (1973). *Ecology* **54**, 315–325.
May, R. M. (1974). "Stability and Complexity in Model Ecosystems" (2nd edn.). Princeton University Press, New Jersey.
May, R. M. (1976). *In* "Theoretical Ecology: Principles and Applications" (R. M. May, ed.), pp. 4–25. Blackwell Scientific, Oxford.
Maynard Smith, J. (1974). "Models in Ecology." Cambridge University Press.
Monod, J. (1950). *Ann. Inst. Pasteur* **79**, 390–401.
Morris, W. D. (1974). "Differential Equations for Engineers and Scientists." McGraw-Hill, London.
Nisbet, R. M. and Gurney, W. S. C. (1982). "Modelling Fluctuating Populations." Wiley, London and New York.
Nyholm, N. (1978). *J. Theor. Biol.* **70**, 415–425.
Paasche, E. (1973). *Mar. Biol.* **19**, 117–126.
Perram, J. W. (1973). *J. Theor. Biol.* **38**, 571–578.
Proper, G. and Garver, J. C. (1966). *Biotechnol. Bioeng.* **8**, 287–296.
Sanchez, D. A. (1968). "Ordinary Differential Equations and Stability Theory: An Introduction." Freeman, San Francisco and London.
Sudo, R., Kobayushi, K. and Aiba, S. (1975). *Biotechnol. Bioeng.* **17**, 167–184.
Tsuchiya, H. M., Drake, J. F., Jost, J. L. and Frederickson, A. G. (1972). *J. Bacteriol.* **110**, 1147–1153.
van den Ende, P. (1973). *Science* **181**, 562–564.
Williams, F. M. (1967). *J. Theor. Biol.* **15**, 190–207.
Williams, F. M. (1971). *In* "Systems Analysis and Simulation in Ecology." (B. C. Patten, ed.). Academic Press, New York and London.

Chapter 4

Catastrophe Theory

P. T. Saunders
Department of Mathematics, Queen Elizabeth College, London UK

Despite its name, catastrophe theory is not confined to a study of disasters. Rather, it is a mathematical way of describing systems which change suddenly. Many biological systems are of this type while most of the properties dealt with in classical physics change smoothly and can be described in terms of differential calculus. It is for this reason that the founder of catastrophe theory, the French mathematician René Thom, considers it to be particularly applicable to biological phenomena. In this chapter the basic elements of catastrophe theory are introduced and examples in which it has been applied to problems in microbiology are presented—*Editorial note.*

1. Introduction

Many changes in nature occur smoothly, like the motion of the planets around the sun or the variation in atmospheric pressure with altitude. Most of applied mathematics has been developed to describe these smooth changes, and of course if something is varying in a totally chaotic fashion it is difficult to see how any mathematical technique could adequately deal with it. Sometimes, however, a system which usually behaves smoothly will undergo a single abrupt change. For example, a wave alters its shape gradually as it approaches the shore; then suddenly it breaks. This is a discontinuity in time, but there are also discontinuities in space: at any instant there is a continuous variation in the properties of the water within the wave but at the surface of the sea there is a definite discontinuity.

We generally try to model smooth variations by using differential equations. Sometimes we find that the solutions of these equations have singularities, points at which the functions or their derivatives become infinite or are not properly defined. These singularities can give us information about the discontinuities; for example, the differential equations which we use to study the continuously varying properties of a fluid which is disturbed by a slowly moving object also predict the shock wave which is associated with supersonic flight.

For the physicist, differential equations are an indispensable tool. In biology, on the other hand, their use is limited, and for obvious reasons.

Biological systems are typically very complex, so we cannot hope to be able to model them with one or two differential equations. Nor are the observations of the same quantitative precision that physicists take almost for granted. In many cases the data are only qualitative, and if there are discontinuities they may be the only repeatable observations: the thickness of the layers of different types of tissue in a human arm will vary from individual to individual, but the sequence of types will always be the same.

Now suppose that we are interested in a system which we believe can be modelled by differential equations, but that we do not know what the equations are. Obviously we cannot hope to determine the continuous behaviour, but it turns out that we can still predict quite a bit about the discontinuities. If we are prepared to make a few quite reasonable assumptions about the system, then catastrophe theory (Thom, 1972) can provide us with a considerable amount of information concerning where and under what circumstances sudden changes can occur.

This rather vague statement of the result will become clearer and more precise as we go along, but at this stage there are two points which ought perhaps to be made. First, catastrophe theory has no necessary connection with disasters, although it has been applied to the study of the collapse of box-girder bridges and the capsizing of ships. The word "catastrophe" is to be taken in its original sense of a sudden and important alteration in the course of events. In the nineteenth century, the term "catastrophists" was used to describe those who thought that certain phenomena in nature had arisen through sudden and violent disturbances, with no implication of calamity. The reader is also assured that we are not claiming to be able to get something for nothing, that catastrophe theory can allow us to get more information out of our analysis than we put in in the form of observations. A knowledge of the locations of the discontinuities of a system represents only a minute fraction of the total information we can have about a system. It may be the most interesting information, but that is a different question altogether.

2. The Zeeman Catastrophe Machine

The easiest way to begin to learn about catastrophe theory is to build a device called a "catastrophe machine" (Fig. 1), which was invented by Zeeman (1972) for the purpose. All that is required in the way of materials are two rubber bands, which should be approximately the same length, three drawing pins, some cardboard and a piece of scrap wood to serve as the base. Cut the cardboard into a disc whose diameter is about equal to the length of the rubber bands, and push one of the drawing pins through

4. CATASTROPHE THEORY

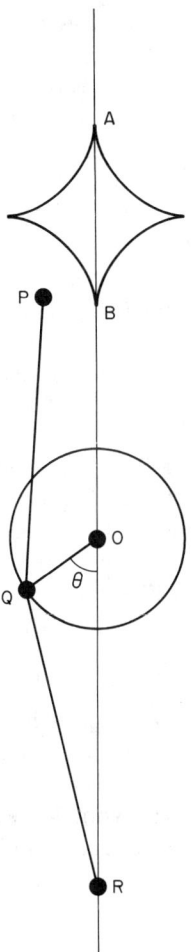

Fig. 1. The catastrophe machine.

this disc at a point Q close to the circumference. With this pin pointing upwards, fasten the disc to the base with a second drawing pin through the centre O. Slip the two rubber bands over the pin at Q, and use the third drawing pin to fasten the other end of one of them to a point R on the base, at a distance from O of about twice the diameter of the disc. Leave the remaining end, P, free.

To operate the machine, move the free end slowly from place to place on the base. You should observe the following behaviour: a small change in the position of P generally produces a small change in the position of Q. Sometimes, however, Q moves suddenly to a quite different position. The positions of P at which the jumps in Q can occur form the perimeter of a curved diamond, but P can cross this perimeter without causing a jump. In particular, if P is moved back and forth at right angles to the line OR then there is only one jump in each direction, and these do not occur at the same place. Also if P is outside the diamond then there is only one possible stable position of Q, whereas if P is inside the diamond there are two, one to the left and one to the right. There is also a third equilibrium position between these two, but it is unstable.

This apparently peculiar behaviour of the catastrophe machine can actually be explained quite easily. Let B be the origin of a cartesian co-ordinate system, let (ξ, η) be the co-ordinates of the free end P, and let θ denote the angle ROQ. Now the potential energy of the system is equal to the work done in stretching the rubber bands. Since this depends only on the positions of the points P and Q it is a function of θ, ξ, η and no other variables; we may write it as $V(\theta, \xi, \eta)$. When we move P to any desired position we fix the values of ξ and η, and the machine responds by adopting that value of θ which minimizes V. Hence we can determine the equilibrium position of the machine by solving for θ the equation

$$dV(\theta, \xi, \eta)/d\theta = 0$$

Naturally the solution of this equation will depend on the values of ξ and η, which reflects the fact that the angle ROQ is different for different positions of P.

It is not difficult to find an expression for $V(\theta, \xi, \eta)$, but the calculations are tedious and we shall not reproduce them here (see Saunders (1980) for details). It turns out to be convenient to transform to new variables (x, u, v) given by

$$x = a_1(\theta + a_2\eta), \quad u = a_3\xi, \quad v = a_4\eta$$

where the a_i are constants whose values depend on the precise dimensions of the machine and on the modulus of elasticity of the rubber bands. We then find that providing P is close to B we may write

$$V(x, u, v) = x^4 + ux^2 + vx$$

The equilibria can then be located by solving the equation

$$4x^3 + 2ux + v = 0$$

Since this is a cubic equation it has either one or three real roots, depending on the values of u and v. Consequently, for some values of these parameters (i.e. for some positions of P) the potential V has a single minimum, whereas for others it has two minima separated by a maximum. We can locate the regions of the u–v plane in which there are one and two minima, respectively, by using the fact that on the boundary between the two regions the function V must have a point of inflection, as one of the minima coalesces with the maximum. So we find the boundary by solving the equations

$$dV/dx = d^2V/dx^2 = 0$$

i.e.

$$4x^3 + 2ux + v = 0 = 12x^2 + 2u$$

The solution of this equation is

$$8u^3 + 27v^2 = 0$$

or in terms of the original variables

$$(8a_3^3)\xi^3 + (27a_4^2)\eta^2 = 0$$

Hence if we draw on the base of the catastrophe machine the cusp-shaped curve this represents, we know that at any point within the cusp V will have two minima, but that outside the cusp it will have only one (Fig. 2). If we now repeat the calculation taking A as the origin instead of B we obtain a second cusp oriented in the opposite direction; taken together these give the curvilinear diamond shown in Fig. 1.

We can now see why the catastrophe machine behaves in the way that it does. To see how the sudden jumps come about, consider Fig. 3. Here are shown two sequences of potentials corresponding to a motion of the free end P in a direction at right angles to the axis of the machine which takes it across the diamond. In each case as P enters the diamond a second equilibrium appears, but this has no immediate effect on the machine. The sudden jump occurs when P leaves the diamond, but only if it leaves by crossing the boundary opposite to the one by which it entered. In that case the equilibrium that the system is in disappears, and the system has no option but to shift abruptly to the other one.

It is easier to see what is going on if we think of the three variables u, v, x (or, equivalently, ξ, η, θ) as cartesian co-ordinates in a three-dimensional *phase space*. The state of the system at any time is represented by a point, called the *phase point*. When the system is at equilibrium it must lie on the *equilibrium surface*, whose equation is $4x^3 + 2ux + v = 0$, and which is shown in Fig. 4. In fact the phase point must always lie on either the top or bottom sheet of the surface, because the middle sheet corresponds to unstable equilibria.

The position of the free end P is represented by a point in the u–v plane,

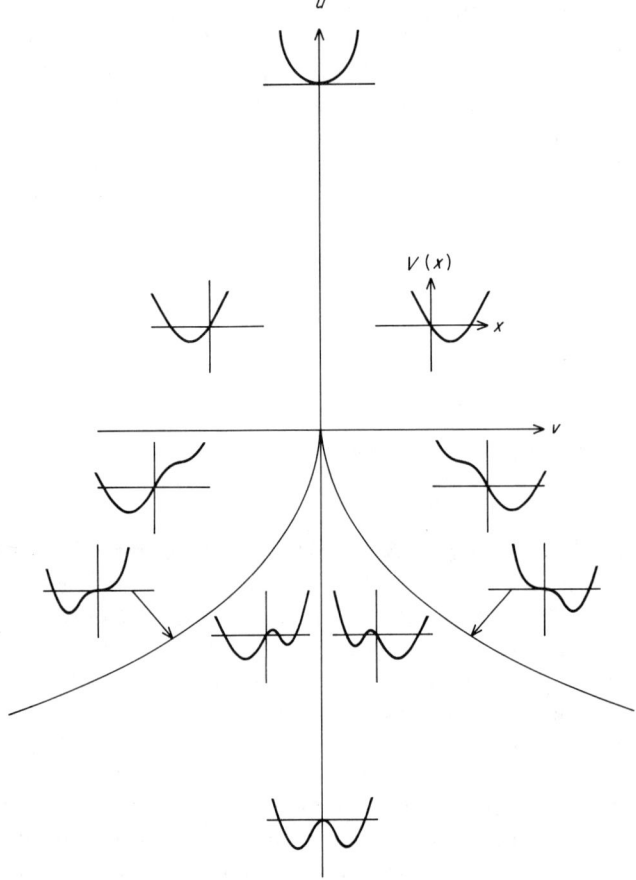

Fig. 2. The control space for the cusp catastrophe, showing the form of $V(x)$ for different values of u and v.

which we call the *control space* for the system. As we move P, the corresponding *control point* traces out a path which we call the *control trajectory*. At the same time, the phase point follows a trajectory which lies in the equilibrium surface and is always directly above the control trajectory. In general, if the control point moves smoothly, so does the phase point. Exceptions to this can occur, however, when the control trajectory crosses the *bifurcation set* $8u^3 + 27v^2 = 0$, which is the projection into the u–v plane of the folds of the equilibrium surface. If the phase point is on the surface which ends at this point (by folding back and

4. CATASTROPHE THEORY

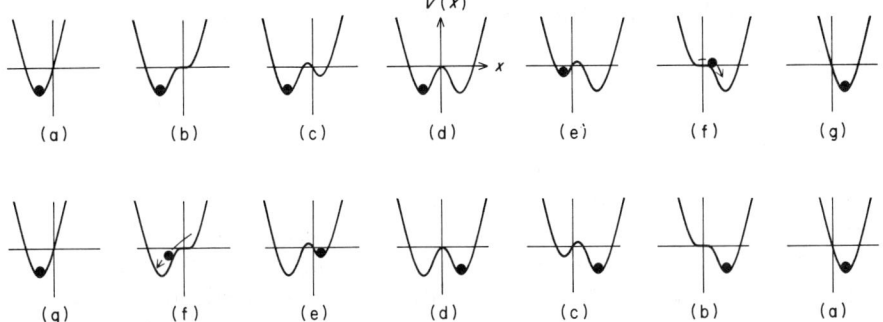

Fig. 3. $V(x)$ for the cusp catastrophe with fixed $u < 0$ and different values of v. This figure illustrates the "perfect delay" convention. The top row is to be read from left to right, the bottom row from right to left, as indicated by the labels.

becoming the middle sheet) then it must jump to the other sheet. There is then a discontinuity in x, and also in θ.

3. Catastrophe Theory

The catastrophe machine illustrates clearly the basic idea of catastrophe theory, that a system with no obvious built-in discontinuities can still

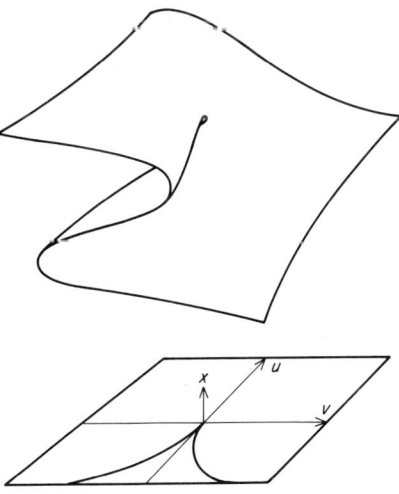

Fig. 4. The canonical cusp catastrophe.

undergo sudden changes on account of the disappearance of stable steady states. It is, however, important to note that while in this example the dynamic of the machine is governed by a potential, this is not necessary. The arguments used in the proof apply equally if we have a Liapounov function so systems which are generally at or near the equilibria of sets of ordinary differential equations are also within the scope of the theory, whether the dynamic is of gradient type or not.

The catastrophe machine is, of course, a particularly simple system, which is why it was chosen as an illustration. It is under the control of only two *control variables*, the co-ordinates of P, and its configuration can be completely specified by a single *state variable*, the angle θ. Moreover, we understand completely the dynamic which governs its behaviour, and we can therefore write down and analyse an explicit expression for the potential.

Suppose, on the other hand, that we are concerned with a large and complex system, whose configuration requires n state variables for a complete description, where n is a large number. For a model of the brain n could be of the order of millions, and indeed we might not even know precisely how large it is. Suppose further that the system is regulated by m control variables, where m is small, no greater than 4 in most cases. This is less of a restriction than it sounds, for we may leave out of account any control variables which have no significant effect on the discontinuities we are studying, and if behaviour which is not smooth depends critically on more than four or five control variables, then we can hardly expect to be able to make much sense of it by any means. Finally, we suppose that the dynamic is governed by a potential, or at least that a Liapounov function exists. This is equivalent to the quasi-steady-state approximation which is often used in biological studies. Catastrophe theory can also apply in other situations, if there is a variational principle, or if some of the commonly encountered partial differential equations (including the wave equation) are involved.

What catastrophe theory then tells us is the following. The number of qualitatively different catastrophes which can occur depends not on the number of state variables, which may be very large, but on the number of control variables, which is generally small. In particular, if the number of control variables does not exceed four, then the configuration of discontinuities is the same as if the system were governed by a potential from the following list of seven canonical forms:

$x^3 + ux$ fold

$x^4 + ux^2 + vx$ cusp

$x^5 + ux^3 + vx^2 + wx$	swallowtail
$x^6 + tx^4 + ux^3 + vx^2 + wx$	butterfly
$x^3 - xy^2 + w(x^2 + y^2) + ux + vy$	elliptic umbilic
$x^3 + y^3 + wxy + ux + vy$	hyperbolic umbilic
$y^4 + x^2y + wx^2 + ty^2 + ux + vy$	parabolic umbilic

Here x and y denote state variables and the other quantities are control variables. It is significant that in none of these potentials are more than two state variables involved.

This is as remarkable a result as it sounds, perhaps even more so. For consider the alternative. To analyse the behaviour of the system by the usual methods we would have to write down a large number—perhaps hundreds of millions—of differential equations. We would have to supply initial or boundary conditions, solve and then try to make sense of the solutions. If, as is likely, we could not obtain analytic solutions, we would have to integrate the equations numerically for many different combinations of initial conditions, in order to have any confidence that we had obtained a complete picture of the behaviour of the system. Even if we knew in advance which were the important state variables we would not be able to restrict our attention to them alone, as we can with catastrophe theory. Coupled differential equations cannot be separated in this way. It is necessary to solve the complete set first and pick out the relevant variables afterwards. In contrast, catastrophe theory can enable us to predict much of the qualitative behaviour of the system without even knowing what the differential equations are, much less solving them. And it does this on the basis of a few assumptions which are surprisingly unrestrictive, and which are probably satisfied by many biological systems.

4. Structural Stability

The key to catastrophe theory is the concept of structural stability. The idea is that in real life nothing is ever exactly repeatable, only approximately so. No matter how carefully we set up the apparatus, the conditions of an experiment are bound to be at least slightly different from those on any previous occasion. So when we say that we expect that experiments must be repeatable, we do not really mean that if we repeat them under precisely the same conditions we will obtain precisely the same results, but rather that if we repeat them under approximately the same conditions we will obtain approximately the same results. A system for which this is true is said to be structurally stable. And if we are going to model it by means of a system of differential equations, then these too should be structurally stable, by which we mean that small alterations in the equations should not have significant effects on the solutions. This is why the well-known Lotka–Volterra equations are unsatisfactory as a model of predator–prey interactions: almost any modification of them, however slight, results in oscillations which are either damped or increasing, rather than persistent.

Now consider a system which can be described by a single state variable, x, and whose dynamic is governed by a potential $V(x)$. Suppose that for one particular set of values of the control variables (i.e. for one particular set of conditions) the potential has the form

$$V(x) = x^4$$

Then there will be a unique equilibrium, at the origin, and it will be stable.

For any other values of the control variables we cannot expect that $V(x)$ will have precisely this form, though if we try to set the control variables as close as possible to the original ones we would expect $V(x)$ to be very close to x^4. So we are led to ask what would be the effect of adding to the potential a term εx^m, where $|\varepsilon|$ is small and m is a positive integer. Will there still be a unique stable equilibrium at, or at least very near, the origin?

It is easily shown that if $m \geq 4$ then the nature of the critical point at the origin is not affected by the extra term, but that if $m < 4$ it can be. For example, if $\alpha < 0$ then no matter how small $|\alpha|$ may be, the function $x^4 + \alpha x^2$ has a maximum at the origin, with two nearby minima, one to either side. So x^4 is structurally unstable, because the addition of a very small term can affect the nature of the critical point at the origin. We can, however, make it structurally stable, by adding to it all terms of degree less than 4:

$$V(x) = x^4 + \alpha x^3 + \beta x^2 + \gamma x + \delta$$

Actually this is more than we need, because any quartic can be written in a form with no cubic or constant term simply by changing the origin of x and V. So the structurally stable "universal unfolding" of x^4 is

$$x^4 + ux^2 + vx$$

which is the cusp catastrophe. The canonical forms of the other catastrophes with one state variable are derived in the same way. The derivations of the three umbilics are less simple, but still not really difficult. The hard part is showing that this set of seven polynomials really does cover all the cases that can arise, and that nothing has been left out. In fact, these polynomials are the first few terms of Taylor series in one or two variables, but since

4. CATASTROPHE THEORY

not every function can be adequately represented by a Taylor series it is by no means obvious that we have included everything. And we have also to prove that two state variables are all that we need consider.

We are, as we have said, concerned with systems which behave smoothly almost all of the time, but which do have some discontinuities. These are isolated and, what is more, occur in a repeatable fashion: if we repeat a sequence of changes of the control variables as nearly as we can, we expect that the sudden jumps will occur in very nearly the same places. We express this mathematically by using a potential which is structurally stable except for certain values of the parameters. These isolated instabilities produce the sudden jumps, while the structural stability of the entire family of potentials ensures the approximate repeatability of experiments which we expect.

5. An Example of the Fold Catastrophe

In this section we analyse the same simple system twice, once by standard techniques and once by catastrophe theory. Our chief aim is to illustrate catastrophe theory by applying it to a problem whose answer is already known, but the reader may find that the second approach is clearer. It also has the great advantage of structural stability, although this is a consideration which is often overlooked when differential equations are used.

Imagine a single-species chemostat culture in which there is growth limitation due to substrate inhibition. If we let x be the biomass concentration, s the substrate, μ the specific growth rate and D the dilution rate, then we may write (cf. Pirt, 1975)

$$\dot{x} = (\mu - D)x$$

$$\mu = \frac{\mu_m K_i s}{s^2 + K_i s + K_i K_s}$$

$$\dot{s} = D(s_r - s) - \frac{\mu x}{Y}$$

where s_r is the concentration of substrate in the input and K_i, K_s and Y are constants. As is customary, a dot denotes differentiation with respect to time.

The steady-state substrate concentration, which we denote by $\bar{s}$, is found by solving the equation $\mu = D$, i.e.

$$\bar{s}^2 = K_i\left(1 - \frac{\mu_m}{D}\right)\bar{s} + K_i K_s = 0$$

Since this is a quadratic equation it has two roots:

$$\tilde{s} = \frac{1}{2}\left\{\left(\frac{\mu_m}{D} - \right)K_i \pm \sqrt{\left[K_i\left(\frac{\mu_m}{D} - 1\right)^2 - 4K_iK_s\right]}\right\} \quad (*)$$

and these are real and distinct, real and equal or complex accordingly as

$$K_i\left(\frac{\mu_m}{D} - 1\right)^2 - 4K_s$$

is greater than, equal to or less than zero. The situation is illustrated in Fig. 5, in which $\tilde{s}$ is plotted as a function of the dilution rate D. We can also show (by the methods of Chapter 2) that where two real steady states

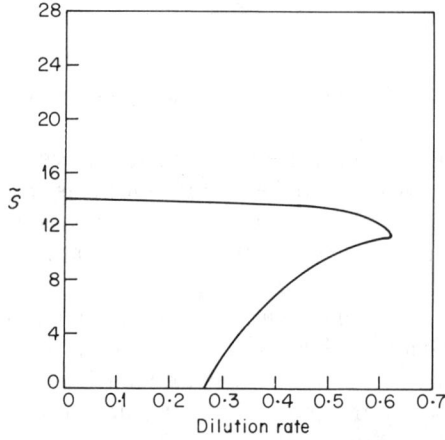

Fig. 5. Equilibrium substrate concentration as a function of dilution rate: $\mu_m = 1.0$; $K_s = 1.0$; $K_i = 10$; $s_r = 28$; $Y = 0.5$. (Modified from Pirt (1975).)

are possible, the one corresponding to the − sign in eqn (*) is stable, but the one corresponding to the + sign is unstable.

Let us now perform the following experiment. We set up a chemostat with the values of the constants equal to those given in the legend to Fig. 5, and we set the dilution rate to 0.3. Naturally all these values are approximate. The substrate concentration will eventually stabilize at an equilibrium value of about 3. We then increase the dilution rate, and we do this sufficiently slowly so as not to disturb the system from equilibrium. The substrate concentration will also increase slowly, but only until the dilution rate is about 0.62. Any increase in D beyond this value will produce not a small change in $\tilde{s}$ but rather a qualitative change in the system, since wash-out will occur.

4. CATASTROPHE THEORY

We have therefore a system which normally responds smoothly to small changes in the single control variable, the dilution rate, but which can exhibit a sudden change, and which we know is almost always at an equilibrium point of a system of ordinary differential equations. Catastrophe theory is therefore applicable, so let us see how it is done.

Because there is only one control variable, the relevant catastrophe must be the fold, whose canonical form is

$$V(x, u) = x^3 + ux$$

The equilibria are found by setting dV/dx equal to zero, i.e. by solving the equation

$$3x^2 + u = 0$$

This equation has real roots if and only if $u \leq 0$, and by differentiating again:

$$d^2V/dx^2 = 6x$$

we see that positive values of x correspond to unstable equilibria, while negative values of x correspond to stable equilibria. When $x = 0$ there is a single equilibrium (or, more correctly, two coincident equilibria) and it is unstable, being a point of inflexion (see Fig. 6).

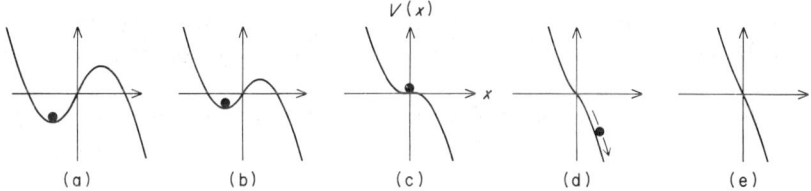

Fig. 6. $V(x)$ for the fold catastrophe for different values of u.

The phase space for this catastrophe is the x–u plane, and this is shown in Fig. 7 in which we have also drawn the equilibrium "surface", which in this case is simply the curve $3x^2 + u = 0$. As with the cusp catastrophe, we can use this diagram to predict the behaviour of any system which is governed by the fold catastrophe, though there are, of course, fewer possibilities.

The control space is the line $x = 0$, and the only control trajectory of interest is the one which begins with u negative and passes through the origin in the positive direction. The phase point moves along the equilibrium "surface" until it too reaches the origin. At this point it can no longer remain on the surface, so it must jump off, just as in the cusp. The difference is that there is no other sheet for it to land on, so the model does not tell us what the eventual configuration of the system will be. This is also illustrated by Fig. 6(c), in which we see there is no local minimum, i.e. no stable equilibrium at all.

Thus the qualitative behaviour predicted by catastrophe theory is precisely the same as that we found by working with the differential equations themselves. Note, however, that catastrophe theory does not tell us at what value of D the sudden change will occur; this is a quantitative prediction and requires more information. In fact, it does not even tell us that a sudden change will occur at all, only that if there is a discontinuity then with only one control variable it must be qualitatively as described by the fold. On the other hand, catastrophe theory guarantees us that even if our estimates of the values of the constants are imprecise or the equations we have written down are not exactly those which correctly describe the system, the sudden changes can only occur in this way. Since we cannot be as certain as we might like about either of these two points, this is a real advantage.

Something else worth noting is that strictly speaking, the predictions of catastrophe theory are valid only in an infinitesimal neighbourhood of the critical point. Yet, as with the catastrophe machine, the result actually turns out to be correct over a much larger range. This is a feature which is not peculiar to catastrophe theory. Many mathematical techniques are customarily used beyond the infinitesimal range with which they are strictly valid: Liapounov's first method, and indeed the Taylor series on which it is based, are examples.

In this example we actually know the relation between the physical

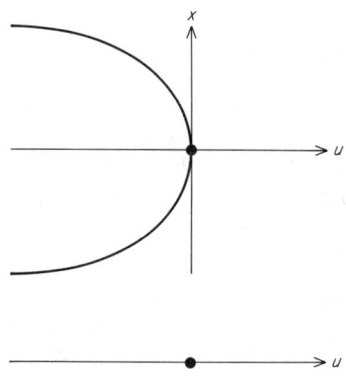

Fig. 7. The canonical fold catastrophe.

variables $\tilde{s}$ and D and the canonical variables x and u; it is given by

$$x = \tilde{s} - \tfrac{1}{2}(\mu_m/D - 1)$$
$$u = \tfrac{3}{4}[K_i^2(\mu_m/D - 1)^2 - 4K_iK_s]$$

In most cases, of course, we do not, which is why the conclusions we draw are generally qualitative. This is illustrated by a comparison of Figs 5 and 7. One can be obtained from the other by a diffeomorphism from the other; this preserves the qualitative features but introduces distortions into lengths.

6. Two Applications of Catastrophe Theory

With many mathematical techniques it is possible to separate the mathematics from the science. Once we have written down the differential equations which we believe represent the behaviour of a system, we can often forget about the original problem completely, until the time comes to compare the predictions with the observations. In many cases the mathematical analysis can be carried out by someone who has comparatively little knowledge of the problem out of which the equations arose.

This is not generally the case with catastrophe theory, especially when it is being applied in biology. There are, in any case, no standard algorithmic techniques; we have instead to be on the lookout for situations in which catastrophe theory might be of use in some problem we are interested in, and then use our own ingenuity to apply it profitably. The two examples which we shall discuss in this section are therefore described in much greater detail than one might expect in an account of a branch of applied mathematics: this is necessary because without following through the problem as a whole it is difficult to see how the theory is really being used.

In studying these examples, there are two points that the reader should note. First, while he may have seen strongly worded claims that catastrophe theory cannot be applied in biology, the examples here are real applications, not mere illustrations. Both produced new results, and both suggested further experiments. Second, while it is no doubt a coincidence that the same organism, *Dictyostelium discoideum*, is involved in both examples, it is probably not a coincidence that two of the earliest real applications of catastrophe theory were in microbiology.

For the benefit of readers who may not be microbiologists, we should explain that *D. discoideum*, a cellular slime mould, is a particularly interesting organism. Its rather unusual life cycle is illustrated in Fig. 8. As long as sufficient nutrient is available it remains in the so-called vegetative

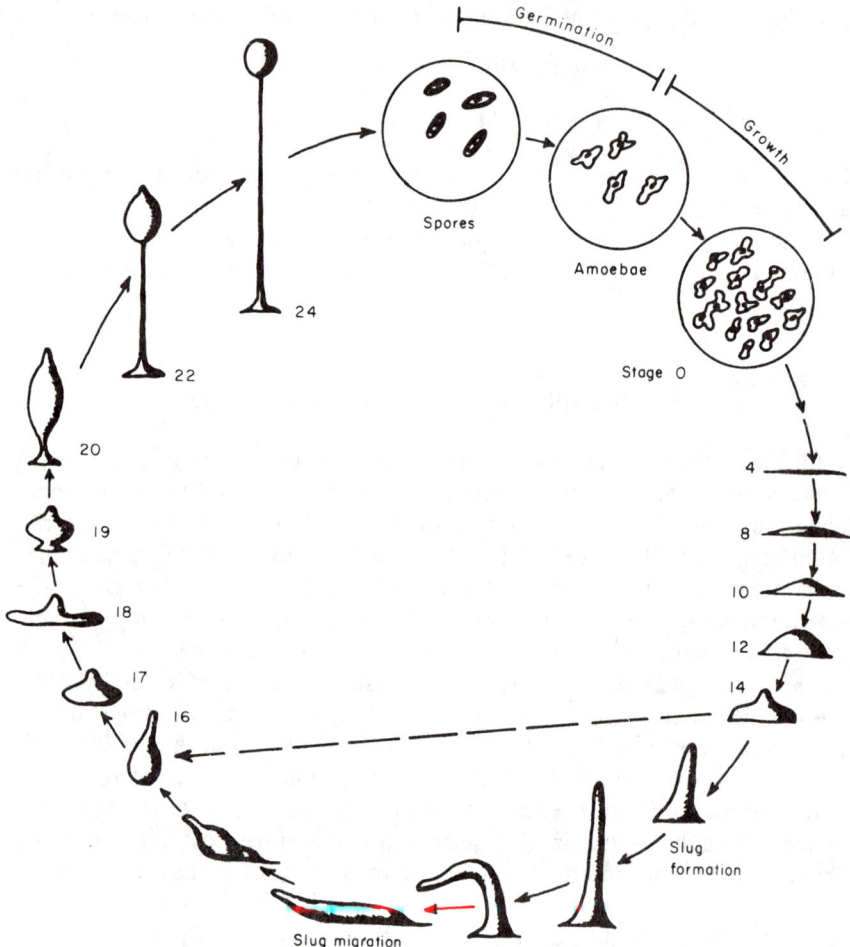

Fig. 8. The life cycle of *Dictyostelium discoideum*. The stages of development are numbered 0–24, and under laboratory conditions these numbers are also the approximate times in hours after the start of development. (Modified from Newell (1971).)

phase, growing and dividing like a typical amoeba. If the nutrient supply fails, however, the cells stop dividing, and if no nutrient is supplied during a further period of about 12 h, they aggregate into a "grex", which migrates, leaving behind a trail of slime from which the organism gets its common name. The grex eventually culminates into a fruiting body, which consists of two quite distinct types of cells, stalk cells and spore cells. The spores float away, and, when they land, germinate to begin the cycle again as individuals in the vegetative phase.

This behaviour is obviously of considerable interest, not only for microbiologists, but also for developmental biologists and evolutionists. The two pieces of research to be described here, which were, incidentally, carried out completely independently of each other, are concerned with two different stages of the cycle. The first deals with the search for the stimulus which first causes the amoebae to divert from the vegetative phase, while the second is an attempt to understand how the rather complicated process of culmination takes place.

6.1. Determination of Critical Variables

This research (Bazin and Saunders, 1978) did not in fact begin as a study into *D. discoideum* at all, but as an investigation into predator–prey dynamics. The original aim was to study the Lotka–Volterra equations

$$\dot{H} = \alpha H - \mu HP$$

$$\dot{P} = \nu HP - \beta P$$

where H and P are the prey and predator population densities, respectively, and the Greek letters all denote constants. These equations were postulated about 60 years ago as an explanation of the oscillations which are often observed in predator–prey systems. Their solutions are in fact periodic, but there are still grounds for doubting that they are a good model. In the first place, the equations are structurally unstable: almost any perturbation of the equations changes the persistent oscillations about a neutrally stable equilibrium to oscillations which either damp fairly rapidly or else increase without limit. Moreover, whenever population oscillations have been studied in detail they have turned out to be better explained as being driven by variations in the environment or, less commonly, as limit cycles arising from some more complicated interaction.

At least part of the reason that there does not appear to be in the literature a well-established example of population oscillations produced by a simple predator–prey interaction may simply be that it is difficult to obtain data which are sufficiently accurate and which cover a sufficiently long interval of time. So it was decided to set up an artificial ecosystem consisting of two micro-organisms, *D. discoideum* feeding on the common gut bacterium, *Escherichia coli*. The expectation was that the situation would be very much improved by the very large numbers of organisms

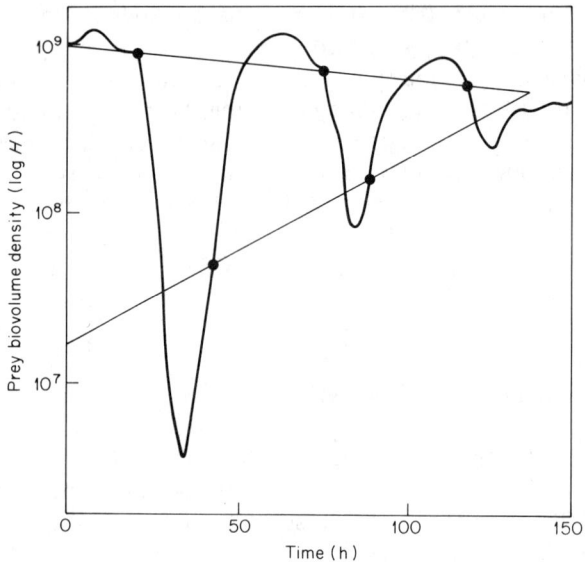

Fig. 9. Prey biovolume. (From Bazin and Saunders (1978).)

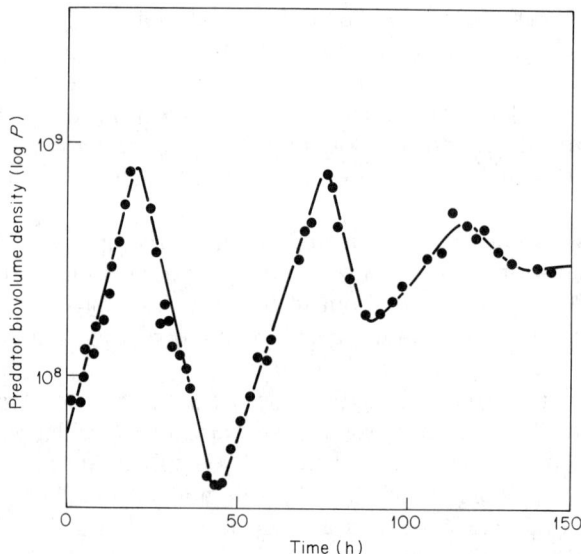

Figure 10. Predator biovolume. The slope of the curve measures the excess of λ over the dilution rate D. (From Bazin and Saunders (1978).)

4. CATASTROPHE THEORY

involved, the short generation times and the relatively well-controlled conditions in a chemostat.

When the experiments were carried out, it turned out that not only did the system not exhibit the persistent oscillations predicted by the Lotka–Volterra equations, it had a number of curious features which seemed difficult to explain using almost any simple model based on differential equations. A typical set of results is shown in Figs 9 and 10, and we can see clearly how λ, the specific growth rate of amoebae, remained constant over quite long periods of time (compared to the generation time of these organisms, which is about 3 h) even though the prey density was changing by a factor of as much as 100. There was then a rapid change to a new specific growth rate, which was in turn maintained for a considerable time despite a large change in prey density. At about 80 h after the start of the experiment all the quantities which were being measured, and their derivatives as well, appeared to be the same as at 30 h, yet at 80 h the populations began to recover, whereas at 30 h they had continued to fall. Another peculiarity, which is not shown in the figures here (see Dent et al., 1976) is that after about 225 h the population densities of both organisms, which had remained approximately constant for some 100 h, moved smoothly to new values, which were maintained until the end of the experiment.

These are the relevant results; now let us try to analyse them. The sudden jumps in λ from one nearly constant value to another indicate that it might be possible to interpret the behaviour of the system in terms of catastrophe theory, with λ as the state variable. The simplest catastrophe on Thom's list that will suffice is the cusp; the fold cannot reproduce the observation that as the experiment continued the jumps became smaller and eventually disappeared altogether. An obvious choice for one of the control variables is the prey biomass density, H. In fact most previous workers have assumed that H is the only factor which significantly influences λ, but having chosen to work in terms of the cusp catastrophe we need a second control variable. We choose t, the time since the start of the experiment; this seems a natural choice, and in any case we know that there have been time-dependent changes in the system.

Note how even at this stage the use of catastrophe theory is affecting the direction of the research. Choosing λ as the state variable immediately shifts the focus of attention from the predator–prey interaction to the response of the amoebae alone. The introduction of time as a control variable is also significant, for it implies that the differential equations which we are assuming describe the behaviour of the system are not of the type that are customarily used in predator–prey dynamics, as these are autonomous.

This choice of control variables makes Fig. 9 a diagram of the control space for the process, with the control trajectory already drawn. We mark the points at which the sudden jumps occur, and sketch a cusp-like curve connecting them. In terms of the canonical variables the equation of the bifurcation set is $27v^2 = 8u^3$, so we have only to find a diffeomorphism (i.e. a smooth transformation of variables) which will carry the equation of the curve into this form. This can be done purely empirically, i.e. by curve fitting, since the functional form of the diffeomorphism is not required.

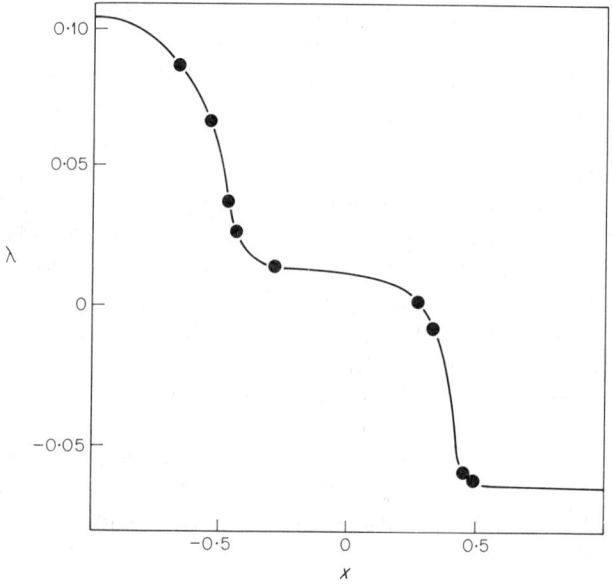

Fig. 11. The empirical diffeomorphism relating the specific growth rate with the canonical variable x. Both the particular data points used and the shape of the curve were chosen on an *ad hoc* basis to achieve the good fit shown in Fig. 10. See text.

We also need a diffeomorphism relating the state variable λ to the canonical state variable x. Now for each data pair (H, t) we have an empirical value of λ. We are also able to use the first diffeomorphism to compute a corresponding pair (u, v) from which we can obtain a value of x by solving the equation $4x^3 + 2ux + v = 0$. This enables us to construct Fig. 11, in which values of λ are plotted against corresponding values of x. In principle we could find some equation which gives us x as a function of λ, but since we do not have to know this equation it is simpler just to transform whatever values of x we need by interpolation. The interpolation procedure assumes that the relation is smooth, and that is all we need.

We are now ready to fit our model to the data. At each point along the control trajectory, we take the values of H and t, use them to compute values of u and v, find x from the equation of the cusp surface and then obtain a value for λ by interpolation from Fig. 11. This gives us predicted values of $\dot{P}/P$ which we integrate numerically to yield a set of predicted values of the predator biomass density P. These are compared with the data in Fig. 10.

The agreement between prediction and observation is obviously good, although what is really significant is only the qualitative picture: the almost straight lines and the decreasing periods and slopes. Once the model reproduced these features correctly we would expect to be able to arrange the close numerical agreement between the observed and predicted values of the slopes by slight *ad hoc* adjustments to the curve sketched in Fig. 11. All the same, the model does appear successful.

There is, however, a problem. At first glance, Fig. 9 may appear to resemble Fig. 2, which is the usual sort of diagram we draw in applications of catastrophe theory, but in fact there is an important difference. The sudden jumps occur not as the trajectory leaves the cusp, but as it enters. This is contrary to the ideas indicated in Fig. 3 and so the model appears to be incorrect.

In fact, it is not as bad as that. While the sequence illustrated in Fig. 3 is the one we normally expect, it is not the only possible one. Some systems do not remain in a local minimum until it disappears. For example, thermodynamic systems tend to move to the global minimum of potential, which results in a sequence like that illustrated in Fig. 12. In terms of the phase-space diagram, this implies that the sudden jumps in both directions occur as the control trajectory crosses a single line in the control space, and the surface of all possible configurations of the system accordingly looks like that shown in Fig. 13. Readers acquainted with classical thermodynamics will recognize the similarity between this figure and the sort of diagram that is often shown in textbooks on the subject, except that it

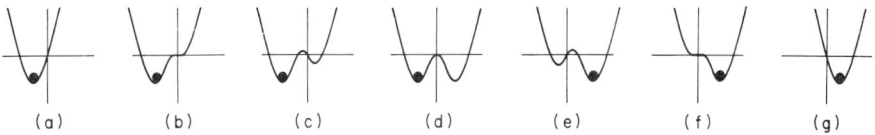

Fig. 12. $V(x)$ for the cusp catastrophe with fixed $u < 0$ and different values of v. This figure illustrates the "Maxwell" convention (cf. Fig. 3).

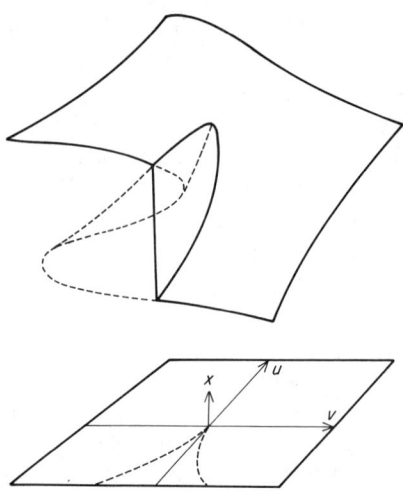

Fig. 13. The cusp catastrophe with the Maxwell convention.

is on its side, as the usual convention in thermodynamics is to take P upwards.

Systems which remain in a local minimum until it disappears are said to obey the "perfect delay" convention; systems which always seek a global

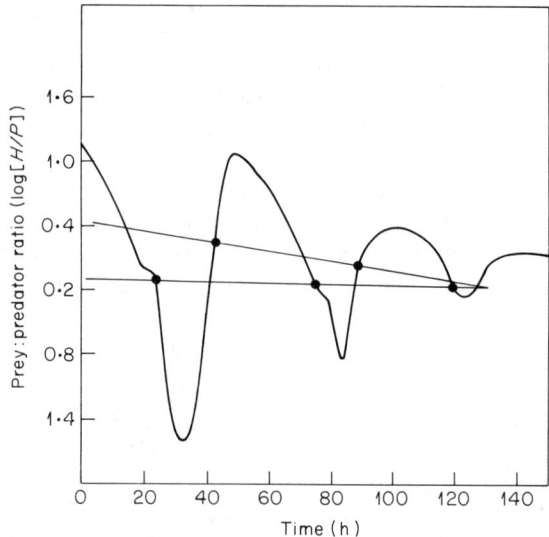

Fig. 14. Ratio of prey to predator biovolume. (From Bazin and Saunders (1979).)

minimum of potential are said to obey the "Maxwell" convention. In the present case, however, we appear to have a system which follows neither of these conventions, since it moves to a new steady state as soon as it appears, even though this state is at a higher potential (see Fig. 3). Now such behaviour is not inexplicable if we are prepared to invoke some additional mechanism which is not in our model. For example, we might suppose that the amoebae are sensitive to the rate of change of the biomass density of their prey. When this derivative is positive the amoebae try to move to the rapid growth mode, but they cannot do this until the bacterial biomass density is great enough for this mode to be available, i.e. until the appearance of the corresponding equilibrium point of the biochemical equations governing the behaviour of the amoebae. As soon as this equilibrium exists, the system moves to it. Conversely, the amoebae react to a falling bacterial biomass by attempting to switch to the slow growth mode, but this too cannot be done until the appropriate equilibrium appears.

Thus it is quite possible to suggest an explanation for the apparently anomalous behaviour of the system. The explanation is, however, somewhat implausible, so before setting out to elucidate the rather complicated mechanism which must be involved, we ought to see if some simpler model can be found. Now the only real freedom we had in constructing the model (as distinct from the choice of diffeomorphisms in the final stages) was in the selection of the control variables, so if we are going to retain a model based on the cusp catastrophe this is the only freedom we have for manoeuvre. Let us consider whether we might have made a better choice.

Most models of microbial predation do suppose that the feeding rate of the organisms depends on the density of the nutrient. There is, however, evidence (Curds and Cockburn, 1971) which tends to suggest that, at least in the case of predatory protozoa, the true critical variable may be the ratio of food to feeders. So we repeat the analysis we have just described, except this time we replace H by H/P. We obtain just as good a fit to the predator data as before, but the difference in the control space (Fig. 14) is striking. The difficulty we described before has vanished, and it appears that the perfect delay convention is applicable.

We have discovered, therefore, that we can fit the data adequately by a model based on a catastrophe with either H or H/P as one of the critical variables. If, however, we choose H, then the mechanism which is involved must be relatively complicated, whereas if we choose H/P the mechanism can be much simpler. We conclude that H/P is much more likely to be the correct choice. Thus catastrophe theory has enabled us to make a useful deduction about the behaviour of the amoebae without the construction of a model in the usual sense of the word.

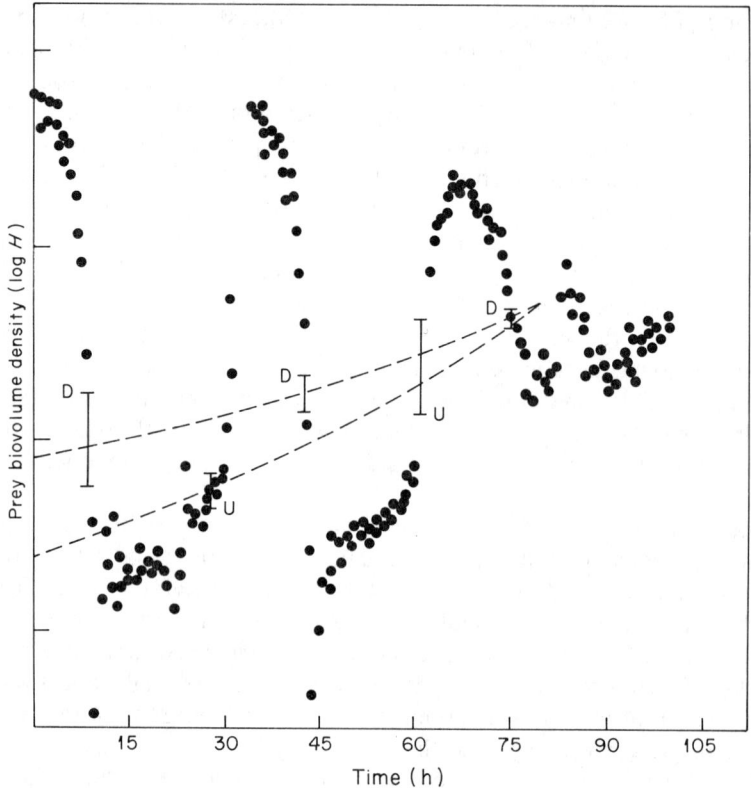

Fig. 15. Prey biovolume. The locations of jumps up and down in the predator specific growth rate are distinguished by letters U and D, respectively, near the error bars. The catastrophe set is indicated by a broken line because its location is known only approximately. Note, however, that no plausible catastrophe set drawn through the error bars would be consistent with the perfect delay convention. (Modified from Bazin and Saunders (1979).)

An important feature of this approach is that it enables us to make use of data which are too noisy to allow us to fit a model based on differential equations. We can see from Figs 9 and 14 that small errors in the values of H or H/P or in the positions of the peaks and troughs of P would not affect our conclusions; the perfect delay convention would still be applicable in the latter but not in the former. Contrast this with the difficulty of estimating the derivatives of the variables, which are well known to be very sensitive to errors in the data. The power of the method is shown to even better advantage by Figs 15 and 16. This work (Owen, 1979) was carried out with a somewhat different aim, and consequently fewer data

4. CATASTROPHE THEORY

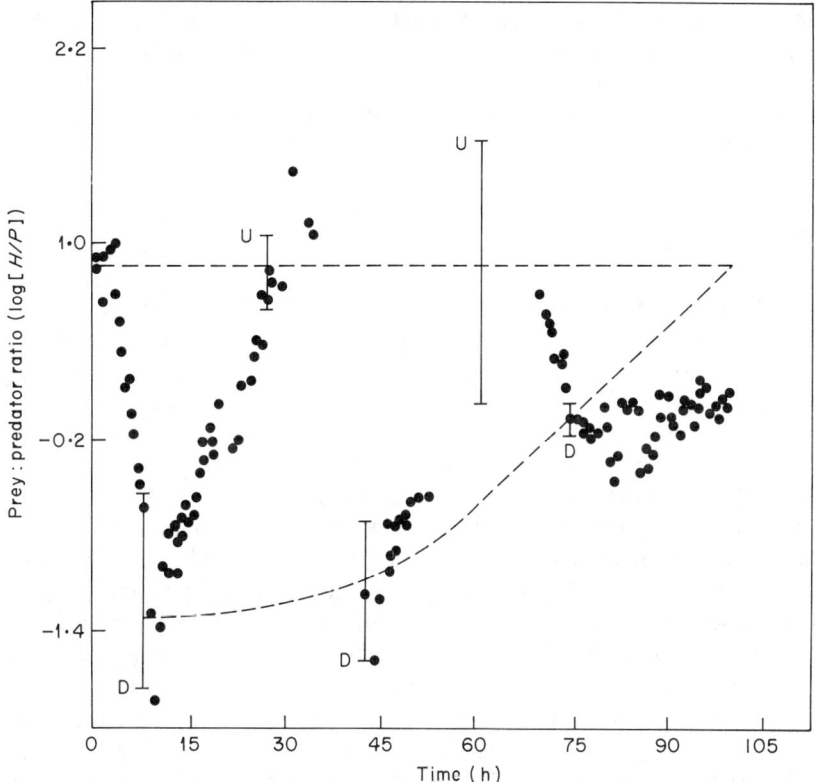

Fig. 16. Ratio of prey to predator biovolume. In contrast to Fig. 15, almost any plausible catastrophe set would be consistent with the perfect delay convention. (Modified from Bazin and Saunders (1979).)

relevant to our problem were obtained. It is doubtful that anything useful could be learned from an attempt to fit a model based on differential equations, but the technique based on catastrophe theory does produce a clear result.

Finally, it is of little use to derive a result by any mathematical technique if it is implausible on biological grounds. So we have to ask how the amoebae could measure H/P, the quantity to which our model predicts they respond. If they were responding to H then they would only have to be sensitive to some substance which is secreted by the bacteria, and an obvious candidate would appear to be folic acid. But if the amoebae respond to H/P then the substance must be one which is both secreted by the bacteria and modified by the amoebae themselves. In particular, if the substance is folic acid, then the amoebae must be modifying it. It is significant that at the same time that Bazin and Saunders (1978) pointed this out, Pan and Wurster (1978) independently reported that *D. discoideum* amoebae do indeed inactivate folic acid, though they did not suggest a purpose for this.

6.2. Culmination

This application is due to Zeeman (1974). The problem is to explain how the grex develops into the fruiting body. This is a rather complicated process, and our immediate task is to use mathematics to suggest a relatively simple way in which it could occur and to propose experiments by which the model could be tested. It seems sensible to test the simplest hypothesis first, and, besides, while nature does not always do things in the simplest way possible, it is generally a good idea to begin by supposing that it does.

Most of the discussion is based on elementary mathematics and physics, but the argument depends on there being a wave of what Zeeman calls "submerging" passing through the grex. This leaves us with the problem of accounting for the wave. We could, of course, simply postulate some mechanism or other which produced this wave at the right time and in the right place, though it is hard to see what the stimulus would be.

A much more attractive hypothesis is that the required wave is what Zeeman calls a "secondary wave". A secondary wave is not a wave at all, in the sense of anything being transmitted, but only a sequence of events taking place independently of each other but each at a fixed time after the passage of a real "primary wave". A good example is the passage of an epidemic across a country. The visible wave of the appearance of the disease is a secondary wave; the primary wave is the wave of infection. The easiest way to distinguish between the two types of wave is that a suitable barrier can stop a primary wave, but not a secondary wave.

The advantage of the hypothesis is that it does away with the need for a special stimulus within the grex; instead we have only to suppose that the submerging occurs in such a way that it takes place not as the primary wave passes through, but some time after. It is not difficult to think of ways in which this could happen. On the other hand, the hypothesis requires the existence of a primary wave. What is more, the primary wave must have some other function as well, for if its only role is to stimulate the cells to submerge, then we are back where we started.

It is at this point that catastrophe theory enters into the discussion. There is nothing in the life cycle of the grex which is an obvious candidate for the primary wave. In fact, up to this point very little beyond aggregation

4. CATASTROPHE THEORY

appears to have taken place. There has, however, been a differentiation of the cells into pre-stalk and pre-spore: outwardly the cells may look the same but those at the front are by now committed to becoming stalk cells and those at the back are committed to becoming spore cells (Raper, 1940).

A frontier has thus appeared in a region which was originally undifferentiated. We now use catastrophe theory to show that it is likely that this frontier moved as it formed, thus providing us with the primary wave we need for our proposed explanation.

To establish this result we need four hypotheses, which as we shall see, are not unduly restrictive. They are:

(1) Homeostasis: Each cell within the grex is in a stable biochemical equilibrium, though this equilibrium may change in time. In other words, we suppose that the usual steady-state approximation is valid.
(2) Continuity: In a newly formed grex the chemical, physical and dynamical conditions vary continuously (if at all) throughout the grex and can therefore be modelled by smooth functions of position.
(3) Differentiation: In a newly formed grex, there is only one type of cell, with at most a smooth variation in properties, but by the time culmination begins there are two distinct types, with a sharp frontier between them. This may not be obvious visually, because the cells have not yet developed into stalk or spore, but we will suppose that there is a substance called a morphogen whose concentration, which we will denote by x, determines what the cell will eventually become. At the beginning x is a continuous function of position, but later a discontinuity appears.
(4) Structural stability: If the conditions are altered slightly, the development of the grex proceeds in much the same way.

The grex is, of course, a three-dimensional object, but we are really only concerned with one spatial direction, that at right angles to the forming frontier. We shall denote distance along the grex in this direction by s, and we shall let t measure the time since the grex was formed. Then we expect x to be a function of s and t. By hypothesis (1) we expect that the value of x corresponding to given values of s and t can be found by finding the equilibrium values of a differential equation, i.e. by locating the minima of a potential (more generally, Liapounov function) $V(x, s, t)$. We take this function to be smooth, and (4) then implies that catastrophe theory may be used. We take x as the state variable and s and t as the control variables.

The next step is to see which of the catastrophes on the list is applicable, and to do this we sketch Fig. 17 which shows what we know about the surface made up of the stationary points of $V(x, s, t)$. Hypothesis (2) tells

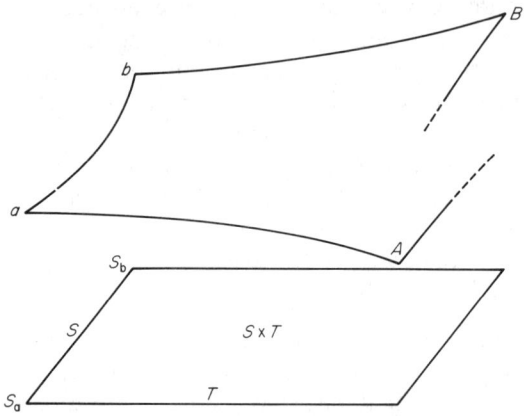

Fig. 17. From Saunders (1980).

us that this surface is continuous along the curve *ab*, and hypothesis (3) tells us that there is a discontinuity along the curve *AB*. This is sufficient to tell us that the simplest catastrophe surface that will fit is the cusp, as the fold clearly will not do. But before we complete the diagram there are a couple of further points we have to consider.

First, it might seem natural to draw the axis of the cusp parallel to the *t*-axis. We should, however, avoid this, as it constitutes a special case, just as when drawing a triangle we ought not to make it isosceles unless we

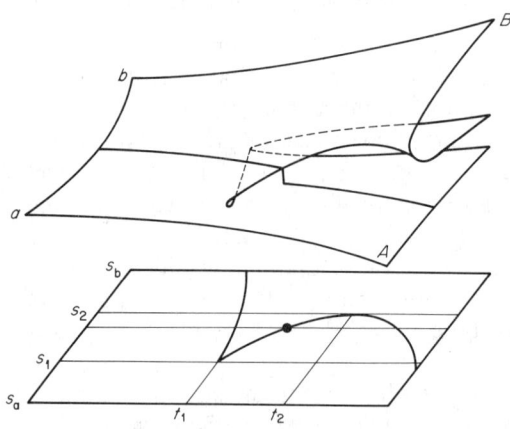

Fig. 18. From Saunders (1980).

know that it definitely should be. Second, we must make the cusp curve back; we shall see shortly that this is necessary if the frontier is to stop. With these extra considerations in mind we can now draw Fig. 18 and use it to discover some of the properties of the model we have constructed.

A line drawn in the surface and parallel to the s-axis gives the concentration of the morphogen at every point in the grex at one particular time t; in particular the line ab represents the situation immediately following aggregation. A line drawn in the surface but parallel to the t-axis represents the concentration of the morphogen at one particular location in the grex throughout the entire period from aggregation to the beginning of culmination. By tracing some of these paths we can see what happens as the differentiation takes place.

The cells at the front of the grex, those in the interval $s_1 \leq s \leq s_a$, all develop smoothly into stalk cells. The cells at the rear of the grex, those in the interval $s_b \leq s \leq s_2$, all develop smoothly into spore cells. The cells in the middle, however, behave differently. They develop smoothly for a while, but they suddenly exhibit an abrupt change in the morphogen concentration, after which they develop smoothly into stalk cells.

If we can continually measure the morphogen concentration throughout the entire length of the grex we will find that originally it will vary continuously from the front to the back. At time t_1, however, a distinct frontier will appear at s_1, though the discontinuity will initially be very small. The frontier will then move through the grex until time t_2, at which point it will stop at s_2. (We can see now why it was necessary to bend the cusp back: had we not required this then the frontier would continue to the back of the grex, and so in the end all the cells would be stalk cells.) Note that the frontier actually reaches its final position in a finite time; it does not approach it asymptotically. After the frontier has stepped, the discontinuity may continue to become greater.

This is the only part of the discussion which depends directly on catastrophe theory, so let us stop to consider exactly what we have accomplished so far. We have shown that if we make certain plausible assumptions about the sort of mechanism that produces the frontier between the stalk cells and the spore cells, then we would expect the frontier to move as it forms. We have not, however, proven that the frontier *must* move, much less that it must move the required distance. And even if the prediction is correct, it does not necessarily follow that the moving frontier has anything to do with the later submerging.

We do, on the other hand, have a hypothesis which we can test. The importance of catastrophe theory is that the hypothesis depends on the movement of the frontier, and without catastrophe theory it is doubtful that we would ever have considered this possibility. Before Zeeman's work it would have seemed natural to suppose that frontiers form in their final position, but we now know that this is not only not necessary, it is probably not even typical.

The next step is clearly to see whether or not the frontier does move, and, if it does, if it moves across the region in which the submerging later occurs. The most obvious way of doing this would be to measure the morphogen concentration directly throughout the grex, but this is not feasible, especially since we do not know the identity of the morphogen. Alternatively, we could cut off different lengths from the fronts of migrating grexes at different times after aggregation (and Zeeman suggests first cutting a single grex lengthwise to provide a control) but so far this experiment appears not to have been performed. The lack of information concerning direct measurements of the movement of the frontier underlines the usefulness of catastrophe theory: if the frontier does move, it does not appear likely that this would have been found experimentally without the strong suggestion from Zeeman's work that it might. In this connection it is worth mentioning that Zeeman also applied his result to the study of the development of an amphibian embryo, and evidence for the movement of a frontier was later found (Elsdale *et al.*, 1976; see also Zeeman, 1978), once the experimenters looked for it.

In the absence of direct evidence, the next best test is to compare the hypothesis with the results of experiments in which portions of the front or back of grexes were removed. Briefly, what was found (see Zeeman (1974) for details) was that the front pieces culminated into stalk only, whereas the back pieces culminated normally, though they took approximately twice as long as usual. If, however, the front pieces were cut off 24 h before culmination, then they too culminated normally.

This behaviour is consistent with the model. The sudden change that occurs is the loss of spore potentiality of the front portion of the grex. If the front piece is cut off, after the primary wave passes then all the cells will develop into stalk cells; in terms of the model they are trapped on the bottom sheet. If, however, the primary wave has not yet passed then the culmination can proceed normally. Equally, since the primary wave does not pass through the back portion of the grex, the cells there retain the capability of becoming either spore or stalk for some time, and so the back portion behaves like the initial grex. Presumably culmination is slower because the gradient is disturbed and has to be re-established, probably by diffusion.

Thus Zeeman's hypothesis is in agreement with the evidence at present available. The remainder of his paper is concerned with the details of culmination and how a very simple model based on a secondary wave of submerging can account for the process; readers who are interested in the details are referred to Zeeman (1974). Zeeman also suggests some experiments that could be performed to test his model, though these do not appear to have been carried out.

In this example, as in the previous one, we have not proposed a mechanism for the process we are discussing. Instead, we have shown that there is a large class of possible mechanisms which could cause a frontier to form, and with the rather unexpected property that they will cause the frontier to move as it forms. The motion of a forming frontier thus becomes something to be expected, not the unlikely event that we might otherwise suppose it to be. As a result, we can feel justified in basing an account of a later process on the hypothesis that the frontier moves, though of course all this will have to be tested by experiment—as indeed must all theoretical predictions, whatever sort of mathematics we use to derive them. And as before, we have a clue which may assist us in the eventual task of determining the mechanism; for example, the existence of a wave suggests that reaction–diffusion equations may be implicated.

7. A Note on Qualitative Results

Even from this brief account of catastrophe theory, it should be clear that there are some important differences between it and most of the other mathematical techniques which are used in biology. The most striking of these is that catastrophe theory, being a part of topology, is essentially a qualitative theory. The state and control variables which appear in the canonical forms are seldom if ever the variables which we measure in the real system. They are, instead, related to the physical variables by smooth transformations. As a result, we can say little about the magnitudes of effects. We can predict that the jumps of the catastrophe machine will occur along two cusped curves, but we can neither write down the equations of these curves nor predict how big the jumps will be.

This may appear unsatisfactory, because we somehow expect mathematics to produce numbers, but there are many situations in which the information we really require is not quantitative. For example, we know that if we use a wiring diagram, or a schematic map such as that of the London Underground, we will learn which components (or stations) are connected directly to each other and which are not, but the actual distances between the symbols may bear little relation to those between the entities they represent. What matters is that the diagram tells us what we want to know; exactly how far it is from Piccadilly Circus to London Bridge is of relatively little concern to the traveller, who is more interested in knowing which train he must take and where he must change.

It is not, however, just a matter of providing information clearly and simply. In many situations, qualitative methods are superior. Consider, for example, the analysis of the catastrophe machine. If we carry this out by the usual methods that a physicist would employ, we can predict the exact locations of the jumps (cf. Poston and Woodcock, 1973)—but only for one particular machine, with given dimensions and a given modulus of elasticity of the rubber bands. But what we really want to show—and this is not in fact proved by the standard method—is that any roughly similar machine, say the one the reader has built for himself, will behave in much the same way. It is this sort of qualitative result which catastrophe theory provides.

Many of the questions that we want answered in biology are essentially qualitative, even though the issue is often obscured by the quantitative nature of the methods we usually employ. As an example, consider the Lotka–Volterra equations:

$$\dot{H} = \alpha H - \mu HP$$
$$\dot{P} = \nu HP - \beta P$$

These are quantitative equations, to be sure, and we can obtain quantitative solutions of them if we wish, but this will not give us quantitative information about the systems they purport to represent. The most they can do is to provide us with some insight into the general behaviour of predator–prey systems, and indeed this was Volterra's aim when he wrote them down. In the same way, most of the work which is being carried out at present using models of this type is intended to produce qualitative results; it is very seldom if ever that quantitative agreement with observations is achieved.

The apparent superiority of the quantitative approach is thus illusory. It produces results in the form of numbers which we do not believe apply to any particular real situation, but from which we draw qualitative conclusions. Qualitative methods, of which catastrophe theory is one—though not the only—example, produce qualitative results directly. What is more, the results are much more robust, for they do not stand or fall with one choice of model. Rutherford may have been right to say that in physics qualitative is just poor quantitative, but in biology there are many situations in which the reverse is true.

8. Conclusion

In this article we have dealt with only one aspect of catastrophe theory. The reader should not, however, be misled into supposing that this is all that there is to the theory. There is already a very wide spectrum of applications ranging from physics to the social sciences, and the ways in which catastrophe theory is applied are as different as the fields in which it is used. For examples, see Thom (1972), Zeeman (1977), Poston and Stewart (1978) and Saunders (1980).

It is, however, in biology that catastrophe theory is likely to make its most significant contribution. This is to a large extent because so many of the most interesting questions in biology are essentially qualitative. But catastrophe theory is also well adapted to biology in another sense. As we have seen, when we apply the theory we have to make some assumptions about the system we are studying, typically that its state at any time can be determined by finding the equilibrium points of a system of ordinary differential equations. Now in physics we are often able actually to write down and solve the equations, in which case we may not need catastrophe theory. In the social sciences, on the other hand, we generally know so little about the underlying dynamic that we are not really in a position to judge whether or not catastrophe theory is applicable, though we may still take it as a working hypothesis. In many biological examples, however, while we cannot write down the equations which govern the system we can be reasonably confident that they are ordinary differential equations, or else the sort of simple partial differential equations to which the theory also applies. Moreover, if an equilibrium assumption is required, this corresponds to the quasi-steady-state assumption which we often make when we study biological systems by other methods. Thus it is in biology in which we expect to have just enough information to allow us to apply catastrophe theory most effectively; in physics we often have too much information, in the social sciences too little.

Catastrophe theory is still very new, and a great deal remains to be done, especially in learning how to apply it. Thom has provided us with a very powerful tool, but with very little in the way of instructions on how it should be used. We shall have to work out the techniques as we go along; there are as yet no algorithms. It is usually not even possible to know in advance what sort of contribution, if any, catastrophe theory is going to make to the solution of any particular problem. The best we can do is to be on the lookout for situations in which the theory might be applicable, and then see whether it—together with our own ingenuity—can tell us anything new. In many cases it may not, but from what we have already seen, we may expect that when catastrophe theory does lead to progress, this will be in directions which might otherwise have remained closed.

References

Bazin, M. J. and Saunders, P. T. (1978). *Nature, Lond.* **275**, 52–54.
Bazin, M. J. and Saunders, P. T. (1979). *In* "Kinetic Logic" (R. Thomas, ed.), pp. 481–501. Springer-Verlag, Berlin.
Curds, C. R. and Cockburn, A. (1971). *J. Gen. Microbiol.* **66**, 95–108
Dent, V., Bazin, M. J. and Saunders, P. T. (1976). *Arch. Microbiol.* **109**, 187–194.
Elsdale, T., Pearson, M. and Whitehead, M. (1976). *J. Embryol. Exp. Morphol.* **35**, 625–635.
Newell, P. C. (1971). *Essays Biochem.* **7**, 87–126.
Owen, B. A. (1979). Ph.D. Thesis, London University.
Pan, P. and Wurster, B. (1978). *J. Bacteriol.* **136**, 955–959.
Pirt, S. J. (1975). "Principles of Microbe and Cell Cultivation." Blackwell, Oxford.
Poston, T. and Stewart, I. N. (1978). "Catastrophe Theory and its Applications." Pitman, London.
Poston, T. and Woodcock, A. E. R. (1973). *Proc. Camb. Phil. Soc.* **74**, 217–226.
Raper, K. B. (1940). *J. Elisha Mitchell Sci. Soc.* **56**, 241–282.
Saunders, P. T. (1980). "An Introduction to Catastrophe Theory." Cambridge University Press.
Thom, R. (1972). "Stabilité Structurelle et Morphogénèse." Benjamin, Reading, Massachusetts (English translation by D. H. Fowler (1975). "Structural Stability and Morphogenesis." Benjamin, Reading, Massachusetts).
Zeeman, E. C. (1972). *In* "Towards a Theoretical Biology" (C. H. Waddington, ed.), Vol. 4, pp. 276–282. Edinburgh University Press.
Zeeman, E. C. (1974). *In* "Some Mathematical Questions in Biology" (S. A. Levin, ed.), Vol. VIII, pp. 69–161. American Mathematical Society, Providence.
Zeeman, E. C. (1977). "Catastrophe Theory." Addison-Wesley, Reading, Massachusetts.
Zeeman, E. C. (1978). *Biosci. Commun.* **4**, 225–240.

Chapter 5

Factor Analysis as an Analytical Method in Microbiology

Maxine A. Holder-Franklin and Lawrence J. Wuest

Environmental Microbiology Research Laboratory, University of Windsor, Windsor, Ontario N9B 3P4, Canada

In Chapter 4 a method was described in which the critical or the most important, variables of an experimental predator–prey system were identified by catastrophe theory. In this chapter the authors use a powerful statistical method, factor analysis, to accomplish the same sort of thing with microbiological data from freshwater environments. In neither case does a unique algorithm for applying these methods emerge. Rather, the specific techniques employed were defined by the results obtained after each step of the two types of analysis—*Editorial note.*

1. Introduction

Many of the scientifically useful mathematical concepts of the nineteenth century did not receive adequate attention from researchers in other fields until the development of the computer. Certainly this is true in applied mathematics and particularly so in multivariate statistics. Few microbiologists have explored the use of mathematics and even fewer the mysteries of factor analysis. Sundman and Gyllenberg (1967), Toerien *et al.* (1969), Howarth and Murray (1969), Toerien (1970) and more recently Rosswall and Kvillner (1978) and Väätänen (1980) have demonstrated its use in studying the complexities of bacterial populations exposed to changing environments. In our studies on population shifts in freshwater river bacteria we have found factor analysis to be a valuable tool (Holder-Franklin *et al.*, 1978; Bell *et al.*, 1982 ab).

Factor analysis combines empirical science with mathematics so that the mathematical constructs are interpreted in the terms of the science being studied. The influences underlying changes in the population structure of aquatic bacteria are complex and interrelated. The application of factor analysis to the problem can assist in elucidating the major influences on the population as well as the correlations between the major influences.

Frequently, in microbiological studies, the wide-ranging database of taxonomic tests is compressed by the classical techniques of taxometrics into a genus and species. The interpretation of a useful body of information is thus restricted. The multitude of characters, if analysed by factor methods, will result in the compartmentalization of the database and also provide new insights into the interpretation of the data.

Used with caution, factor analysis can be an important stage in the examination of ecological data. The development of the methods used in our research on freshwater bacteria will be presented in the following order: an outline of the general concept; a brief description of some of the specific techniques used; and an in-depth description of one ecological study which illustrates the potential and some of the pitfalls at each stage of the calculations.

2. Factor Analysis and the Selection of the Database

2.1. Factor Analysis

Factor analysis is a multivariate statistical technique for extracting the significant dimensions of variation in a study composed of a large number of variables. Variables with common variance can be quantitatively related. The interrelationships of data variables have been determined by their respective scores on a selected group of subjects. The technique manipulates the matrix of intercorrelations of the data variables to elucidate their communality and uniqueness. The variance of the complete data set is condensed into fewer dimensions which are linear constructs of all the variables and which yield quantitatively, the same information about the subjects as the complete database.

The principal factor method as described by Harman (1960) and Comrey (1973), was utilized in our study. The principal factor solution consists of N factors, where N is less than or equal to the total number of characters in the study. Each factor is composed of a series of loadings giving the correlation of each character with that factor. Geometrically, the variable loadings may be viewed as the cosine of the angle between the unit vector representing the variable in hyperspace, and the unit vector representing the factor in hyperspace. The higher the variable loading on a factor, the more closely the factor reflects the variance of that variable. The sum of the squared variable loadings on a given factor is called the eigenvalue of the factor and is a measure of the variance condensed into the dimension represented by the factor. Once a set of factors is found, the factor coordinate frame is rotated so as to produce more scientifically interpretable factors. The rotational method should maximize the variance of the set of squared loadings on the set of factors. Qualitatively, rotation should pro-

5. ANALTYICAL METHOD IN MICROBIOLOGY

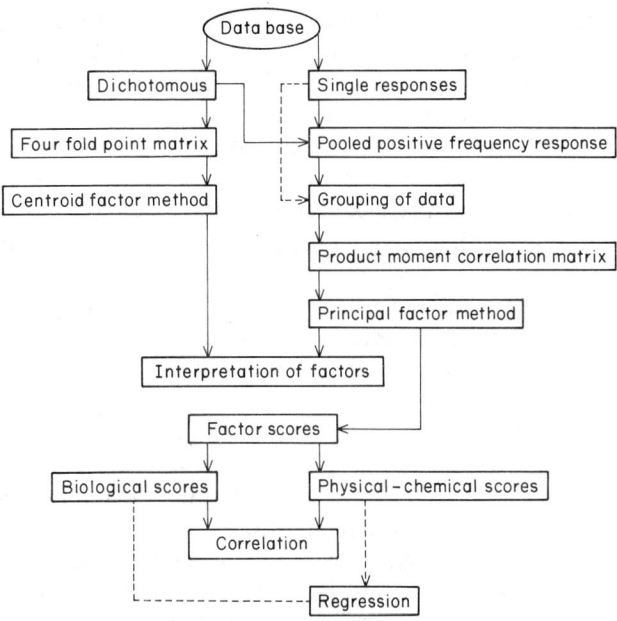

Fig. 1. Flow diagram of procedures to measure microbial responses to environmental change.

duce high and low levels of loadings on any one factor. Variables will tend to weigh highly on one or two factors only, simplifying interpretation.

The factor methods used to study the test responses of bacteria and the physico-chemical parameters of the water are shown in a flow diagram (Fig. 1).

Two major approaches have been developed. In the beginning of the development of factor analysis for population shift studies, the first matrix was formed from the dichotomous test responses by using the fourfold point coefficient, the centroid factor method and hand rotation to a positive manifold. This method was used in our first report (Holder-Franklin et al., 1978) where the test responses of each of the individual isolates from each sample were analysed.

The second approach was to determine the positive feature frequency responses of an appropriate experimental group. The selection of the group is based on an informative parameter such as time, location and the clusters which have been predetermined by numerical taxonomy. In the latter case, the test responses of the selected isolates form a new matrix based on the positive responses.

The dotted line connecting Database to Grouping of Data signifies that data can be factor analysed without obtaining the pooled positive frequency response and, in fact, the physical/chemical data was analysed after grouping by station (synonymous with sample).

The factors are interpreted by the weighting or loading of each variable (test response or physical/chemical data) on that factor. The converse of the interpretation is often reflected in negative loadings.

In order to relate experimental groups to factors within the analysis and to factors in other analyses within the realm of the investigation, it is necessary to obtain a factor score. The data for each variable is analysed to provide an estimate of the factor score on any one factor for an individual, group or station depending on the previous experimental group selection of the data.

The factor scores are computed by various methods (Comrey, 1973). However, most computer packages have standardized to the mathematically exact regression method. Variables with high loadings would have a greater effect on estimating the factor scores.

The flow diagram of the factor method indicates that factor scores of the various data sets, biological and environmental should be examined for possible correlations. This can be accomplished with Pearson product moment correlation or regression analysis.

The method then involves the construction of an ever diminishing set of variables as the first stage in the attempt to elucidate one facet of the ecology of river bacteria, i.e. the response of the free-floating population to environmental change.

2.2. Numerical Taxonomy; the Selection of the Database

The isolates were tested by 226 nutritional and physiological tests. The selection of these tests has been described by Cormier (1978). The species were clustered by the unweighted pair group with arithmetic averaging (UPGMA) of the Jaccard association coefficient of Sneath and Sokal (1962). The entire database before clustering was utilized in the factor analysis.

3. Population Shifts in Heterotrophic Bacteria

Population shifts in the aerobic, heterotrophic, free-floating bacteria of a small freshwater river exposed to a wide range of seasonal changes have been studied by factor analysis. The two sets of data consisted of the physiological responses of the bacteria and 19 environmental parameters.

5. ANALTYICAL METHOD IN MICROBIOLOGY

The particular combination of analyses used in this study are unique and contribute to a new methodology in the examination of bacterial population shifts.

In a flowing river, the chemical and physical parameters change hourly and seasonally suggesting that the bacterial populations may respond to these changes. The total heterotrophic bacterial population at any one time contains a large number of strains occupying many ecological niches. A facet of this complex problem was resolved by correlating the activities of the bacteria with the changing physical and chemical parameters of the environment.

The physiological responses of the freshly isolated bacteria formed the biological data matrix and 19 physico-chemical parameters were taken to represent the environmental influences. Using similar mathematical methods on the bacterial data matrix and the physico-chemical matrix, the significant factors of each were extracted. Each of the two matrices was analysed independently. The physico-chemical data were not pooled into a positive frequency response (Fig. 1).

The factor scores from each analysis were cross-correlated providing a direct mathematical correlation of the functions of a random population of heterotrophic bacteria with a matrix of environmental parameters simultaneously measured.

We tested this methodology by measuring population shifts of heterotrophic bacteria in the Meduxnekeag, a small, rocky bottomed, international river with a watershed in the State of Maine and the Province of New Brunswick. The effluent into this river was largely woodland and potato farming run-off and some treated sewage from Houlton, Maine, U.S.A. A starch processing plant was in intermittent operation when the samples were taken, in 1975–76. All of these sources were many miles upstream from the sampling point. The samples were collected at the Naquadat water quality monitor of the Department of the Environment, Canada. Several chemical and physical parameters were simultaneously monitored by probes connected to a computer terminal, others were performed in the laboratory of the Water Quality Branch, Department of the Environment, Moncton, New Brunswick.

In summary, the analysis demonstrated that oxygen levels in the water were found to be significantly correlated to the metabolic reactions of bacterial isolates; these included respiratory metabolism, nitrification and lipid catabolism. Salt tolerance of the bacteria was shown to be positively correlated with the specific conductance and the ion concentration of the water. Nitrate reduction was shown to be correlated with high CO_2 concentration and lower O_2 levels. The method used to arrive at these conclusions is described.

Table 1. Physical and chemical data for each sample.

Test name	September				October				March				July			
	12.00	18.00	24.00	06.00	12.00	18.00	24.00	06.00	12.00	18.00	24.00	06.00	12.00	18.00	24.00	06.00
Conductance (μS cm^{-1})[a]	247	243	243	238	216	212	214	214	159	155	156	161	146	146	146	153
pH	8.1	8.5	8.6	8.0	9.1	9.3	9.2	8.7	8.1	8.2	8.1	8.2	6.8	7.5	8.2	7.2
Water temprature (°C)	12.4	12.2	11.3	9.6	10.1	12.0	10.7	10.8	0.1	0.0	0.0	0.1	23.2	24.7	28.0	25.3
DO_2 (mg l^{-1})	10.1	9.9	10.2	11.6	16.0	13.7	15.2	15.0	13.9	13.8	14.0	14.1	8.6	9.3	8.0	6.2
DO_2% saturation	94	92	92	101	140	120	140	133	95	94	96	96	100	110	100	75
Carbon (mg l^{-1})	4.0	2.8	8.6	7.4	5.6	6.9	7.6	4.3	4.8	4.8	4.8	4.8	9.4	10.5	9.8	11.3
Nitrogen (mg l^{-1})	0.17	0.17	0.17	0.17	0.17	0.17	0.17	0.12	0.10	0.10	0.10	0.50	0.45	0.40	0.40	0.40
Phosphorous (mg l^{-1})	0.015	0.010	0.015	0.015	0.015	0.02	0.025	0.015	0.005	0.005	0.001	0.005	0.029	0.034	0.029	0.027
Turbidity (mg l^{-1})	1.3	1.4	1.4	1.4	1.7	1.7	1.7	1.7	1.2	1.5	1.5	1.5	2.0	1.5	1.7	2.3
Magnesium (mg l^{-1})	3.2	3.7	3.7	3.7	3.1	3.3	3.5	3.8	3.0	2.5	2.5	3.0	2.2	2.0	2.5	2.5
Sulphate (mg l^{-1})	21	21	21	21	18	18	18	19	15	15	13	13	7	7	7	7
Potassium (mg l^{-1})	1.2	1.3	1.3	1.3	1.0	1.0	1.0	1.0	0.6	0.6	0.6	0.6	0.8	0.5	0.6	0.7
Sodium (mg l^{-1})	4.6	4.6	4.6	4.6	3.3	3.5	3.7	3.9	3.3	2.5	3.7	2.9	2.9	2.7	3.1	2.9
Calcium (mg l^{-1})	33	34	34	34	32	33	33	34	25	24	19	23	24	23	24	24
Nitrogen (mg l^{-1})	0.10	0.10	0.10	0.10	0.02	0.02	0.02	0.02	0.02	0.02	0.02	0.02	0.02	0.02	0.00	0.00
Chloride (mg l^{-1})	8.1	8.1	8.1	8.1	6.7	6.7	6.7	6.8	5.9	5.4	6.6	5.2	4.7	4.4	4.8	4.8
NO_3–NO_2 Nitrogen (mg l^{-1})	0.1	0.1	0.1	0.1	0.1	0.1	0.0	0.0	0.7	0.7	0.7	0.7	0.2	0.2	0.2	0.2
Bicarbonate (mg l^{-1})	114	114	114	114	100	100	100	100	68	66	68	68	62	62	69	68
CO_2 (mg l^{-1})	1.5	1.5	1.5	1.5	0.1	0.1	0.1	0.3	1.2	1.3	1.3	1.2	12.7	3.3	0.8	6.8

[a] MicroSiemens per centimetre units for specific conductance.

3.1. Materials and Methods

3.1.1. Sampling

The dichotomous database of numerical taxonomy obtained from the bacterial isolates from four diurnal river water samplings performed in September, October, March and July, a total of 16 samples, was analysed for predominant influences by principal factor analysis. In addition, the same analysis was performed on chemical and physical data obtained from the same samples (Table 1).

Individual bacterial isolates (a total of 1600) obtained from water samples were scored on a battery of 226 dichotomous characters or tests which included the utilization of 139 substrates and a variety of taxonomic bacteriological tests commonly used to identify or classify the genera found in this stream. The taxonomic method has been described in detail in a previous report (Holder-Frenklin, 1981), the tests are listed in Table 2. The data is available on file by request from the authors. The tests were all related to physiological or nutritional functions of the bacteria.

3.1.2. Microbiological Data Manipulation

As stated previously, individual isolates from the Naquadat station of the Meduxnekeag River were scored on a battery of dichotomous characters. The results for individual isolates were pooled into a positive response frequency for each station on each character as suggested by Sundman (1970, 1973) for soil bacteria in ecological studies. The primary data and positive response frequencies of each character were examined across all isolates and stations for invariate characters or characters with insufficient range and for characters with incomplete data and distinctly non-normal distributions. All deficient characters were removed to avoid undefined or spurious correlation coefficients as discussed by Comrey (1973). The reduced character list of positive response frequencies, now 173, was then utilized to compute the intercorrelation matrix giving the product moment correlation coefficient of each character with every other. Tests eliminated are listed in Table 3.

3.1.3. Naquadat Data Analysis

Individual station samples were monitored for pH, temperature, turbidity specific conductance, dissolved oxygen (DO), and DO% by the Naquadat automatic monitor. Laboratory analyses for O_2, C, N, P, Mg, SO_4, K, Na, Ca, NH_3, Cl, NO_3, NO_2, HCO_3, CO_2 were performed on each sample and

Table 2. Itemized

Factor MX 1		Factor MX 2	Factor MX 3	Factor MX 4
Interpretation Oxidative metabolism Nutritional versatility		Interpretation Salt tolerance Oxidative metabolism	Interpretation Lipid catabolism	Interpretation Cold temperature Polymer production
Eigenvalue 74.5 percentage of trace 40.5		Eigenvalue 35.5 percentage of trace 19.3	Eigenvalue 20.4 percentage of trace 11.0	Eigenvalue 14.7 percentage of trace 8.0
Positive loadings +0.500–0.980		Positive loadings +0.500–0.999	Positive loadings +0.500–0.999	Positive loadings +0.500–0.895
Substrates acetamide adonitol β-alanine creatine amylamine D-arabinose D-arabitol 2-amino-2-methyl-L-propanol analine dulcitol dihydroxyacetone erythritol ethylene glycol L-fucose N-glycylgycine hydroxy-L-proline inulin L-isoleucine 2-ketogluconate lactose L-leucine L-lysine L-lyxose D-melizitose L-methionine methanol oxalate phenol phenyl-1-2-ethandiol n-propanol putrescine urea isopropanol L-cystine L-alanine allantoin betaine ethanolamine ethanol D-fucose D-fructose DL-glycerate glycine L-histidine kynurenate laurate D-mannitol meso-inositol α-methyl-D-glucoside L-ornithine propylene glycol	phenethyl alcohol L-rhamnose sorbitol N-acetylglucosamine aesculin A.-agar L-arabinose DL-arginine arbutin n-butanol DL-lactate D-mannose D-melibiose L-threonine acetate cellobiose D-galactose D-glucose L-phenylalanine L-proline D-raffinose salicin sucrose L-tyrosine valine gluconate glycerol succinate L-tryptophan *Other tests* oxidase arginine dihydrolase 0.1% crystal violet fluorescent pigment growth on 0.1% basic fuchsin bile and brilliant green A*. 0.1% methyl red 0.1% pyronine 0.1% teepol nitrate reduction phosphatase tributyrate hydrolysis tween 20 hydrolysis *Negative loadings* −0.500 to −0.940 glucose fermentation glucose oxidation growth Na₂S₂O₃ H₂S on cysteine H₂S on Na₂S₂O₃ nitrate reduction	Substrates cis-aconitate L-ascorbic acid L-aspartate citrate D-ribose saccharate succinate aesculin fumarate gluconate maltose pyruvate valine *Other tests* catalase oxidase casein hydrolysis gelatin hydrolysis starch utilization 0.1% teepol 0.1% thionine sodium selenite 0.5% NaCl 1.5% NaCl 3.0% NaCl 5.0% NaCl 7.0% NaCl 10.0% NaCl 1.0% bile salts 0.1% bromthymol red 0.1% phenol 0.008% KCN 10% ethanol triple sugar iron A. brain heart infusion A. trypticase soy A. growth at pH 5.0 growth at pH 10 growth at 37°C growth at 20°C phenylalanine A. Pseudomonas A. trypticase glucose extract A. Simmons citrate A. SIM A. growth at 41°C growth at 43°C MacConkey A. glucose oxidation indol production *Negative loadings* −0.516–H₂S production on Na₂S₂O₃	Substrates adipate L-phenylalanine p-phenylethylamine sebacate L-tryptophan hippurate L-proline L-tyrosine L-threonine valine *Other tests* tween 20 hydrolysis tween 60 hydrolysis tween 80 hydrolysis tween 85 hydrolysis 1.0% tetrazolium salt brilliant green bile A. 0.1% pyronine growth at pH 4.0 Simmons citrate A. growth on Levine methylene blue A. 0.1% basic fuchsin 0.1% methyl red *Negative loadings* −0.500 to −0.810 Substrates caproate caprylate arbutin *Other tests* phosphatase starch hydrolysis growth at 43°C growth at 41°C Voges-Proskauer H₂S on Na₂S₂O₃	Substrates adipate L-aspartate 3,4-dihydroxybenzoate fumarate L-glutamate hippurate poly-β-hydroxybutyrate isovalerate 2-ketogluconate mucate pimelate *Other tests* tween 60 hydrolysis levan production growth at 4°C gas from glucose indol

* A. agar

bacterial factors

Factor MX 5	Factor MX 6	Factor MX 7	Factor MX 8
Interpretation Nitrate reduction	Interpretation Nitrification	Interpretation Growth on lower nutrient levels	Interpretation Oxidative/fluorescence
Eigenvalue 9.1 percentage of trace 4.9	Eigenvalue 3.5 percentage of trace 1.9	Eigenvalue 3.1 percentage of trace 1.7	Eigenvalue 6.0 percentage of trace 3.3
Positive loadings +0.500–0.999	Negative loadings −0.802 denitrification	Positive loadings +0.500–0.755	Positive loadings +0.500–0.713
Other tests nitrate reduction	Other tests	Other tests Growth on 1/10 strength brain heart infusion A. Growth on 1/10 strength trypticase soy A.	Substrates Spermidine PO_4
Negative loadings −0.500 to −0.829			Other tests oxidase gelatine hydrolysis King B-fluorescent pigment 1% tetrazolium salt tributyrate hydrolysis nitrite reduction
Substrates cellobiose malate pyruvate raffinosie salicin trehalose tributyrin transaconite D-xylose starch			

Table 3. Tests eliminated because of very low feature frequencies.

Substrate utilization tests	Other tests
δ-aminovalerate	tween 40 hydrolysis
benzoate	pectin hydrolysis
benzylamine	anaerobic growth, 20°C
2–3 butylene glycol	King A phenazine pigment
butyrate	poly-β-hydroxybutyrate
n-butanol	accumulation
carragheenan	growth on 0.04% tellurite
DL-citrulline	growth on 15.0% NaCl
DL-carnitine HCl	1.0% phenol
deoxycholate	growth on cysteine naedium
ferulic acid	growth on $Na_2S_2O_3$
D-galacturonate	growth on TCBS[a]
glutarate	growth on Pseudosel agar
guanine	growth on Pseudo agar
m-hydroxybenzoate	methyl red test
o-hydroxybenzoate	
laevulinate	
maleate	
malonate	
DL-methyl glutamate	
pantothenate	
pelargonate	
propionate	
pyridoxine	
sarcosine	
L-sorbose	
suberate	
DL-serine	
D-tartate	
L-tartrate	
taurine	
tertiary amyl alcohol	
tertiary butanol	
thiamine HCl	
tryptamine	
uracil	
valerate	
xanthine	
xylene	
itaconitate	
carboxymethylcellulose	
nicotinate	

[a] Thio sulfate, citrate, bile salts, sucrose, agar.

the actual data used to calculate the intercorrelation matrix among the physical-chemical parameters.

3.1.4. Factor Analysis Detailed Methodology

Original communality estimates in the principal factor analysis were taken as the largest off-diagonal element of each character in the correlation matrix. The factor solution was iterated by re-estimating the communalities after each iteration until convergence. Re-estimation of the communalities was made as the sum of the squared loadings of each variable in the preceding factor solution. The convergence criterion was 10^{-5}. The principal factors were rotated using the varimax technique as described by Harman (1960).

Computer manipulation of the primary data matrix was carried out under the control of NTSYS, Numerical Taxonomy System of Multivariate Statistical Programs of Rohlf et al. (1971). The relevant NTSYS routines used were:

(1) SIMINTV—computes the correlation coefficient matrix.
(2) FACTOR—general factor extraction.
(3) ROTATE—rotates a factor matrix to simple structure using various criteria.
(4) ALGEBRA—performs matrix multiplication.

Rotated factors were examined for high loadings to determine the collection of characters comprising each factor.

Factor scores were computed for each station by method 3 of Comrey giving the station factor score as the sum of products of character scores and character weights given by,

$$S_{ij} = \sum_{k=1}^{N} C_{ik} F_{kj} \quad (1)$$

where S_{ij} is the score of station i on factor j, N is the total number of characters, C_{ik} is the positive frequency of station i on character k and F_{kj} is the weight of character k on factor j.

Primary data was standardized to avoid unequal character weights due to an unequal variable range and varied measurement scales. Factor scores were standardized to a mean of 0 and a standard deviation of 1 for convenience of presentation.

The significance of the differences between factor scores for individual stations was tested by analysis of variance. Stations were broken down into four random subgroups of 16 isolates each.

Positive response frequencies and factor scores on the eight selected factors were calculated for each subgroup using eqn (1). Using the four subgroups of each of the 16 stations, a between stations/within stations variance ratio was calculated for each factor, testing the null hypothesis

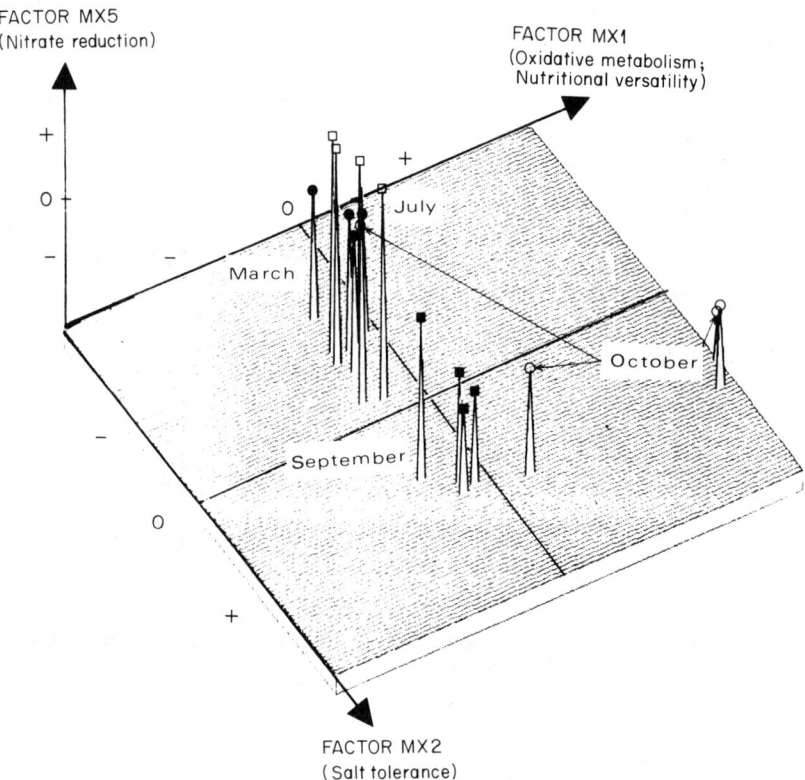

Fig. 2. Spire diagram of 16 biological stations projected in factor space. Each spire represents one station. Factors MX 1, 2 and 5 are utilized. Diagram plotted by computer. ■, September; ○, October; ●, March; □, July.

that only chance variation occurred in the factor scores. The within-station variance of each factor was also utilized to calculate the between-station variance which would yield a significance level of 0.01 for rejection of the null hypothesis. Two standard deviations of this hypothetical population was then used as the level of significant variation on that factor.

5. ANALTYICAL METHOD IN MICROBIOLOGY

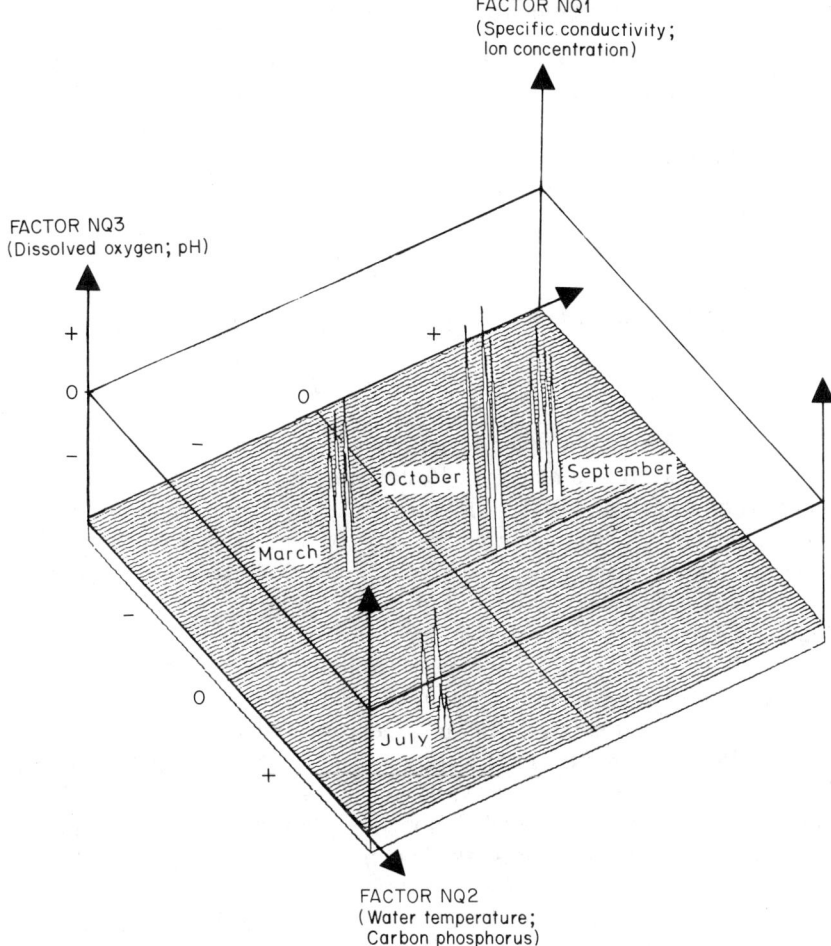

Fig. 3. Spire diagram of 16 physico-chemical stations projected in NQ 1, 2 and 3 factor space.

3.1.5. Factor Graphics

Factor scores were condensed graphically into a two-dimensional projection of the three most significant factor dimensions using the SYMVU program of the Harvard University Laboratory for Computer Graphics and Spatial Analysis (Figs 2 and 3).

3.1.6. *Correlation of Biological and Environmental Factors*

The two sets of factors obtained from the microbiological analysis and chemical-physical analysis, respectively, were analysed for overlap of variance. Pearson product moment correlation analysis of the Statistical Package for the Social Sciences (SPSS) was utilized (Table 7) to demonstrate the interrelationships of the variables.

3.2. Factor Analytic Results

3.2.1. *Microbiological Data*

The positive response frequency data matrix of 16 stations and 226 characters was reduced to a matrix of 16 stations and 173 characters by refinement criteria discussed under methods. Principal factor analysis was performed on 173 × 173 intercorrelation matrix computed from the positive response matrix. Factors were extracted until the eigenvalue of a factor fell below 1.0. Fourteen extracted factors accounted for 99% of the total character variance. Upon rotation, eight factors emerged with eigenvalues greater than 2% of the total common variance. Two percent was taken as an arbitrary limit of chance communality. Table 2 itemizes the various factors at varying loading levels in order to provide sufficient character information for biological interpretation. The eight factors shown account for 94% of the total common variance and 93% of the total variance. Station scores on the rotated factors are shown in Table 6. Prior to calculation of the factor scores, all characters were standardized to a mean of 0.0 and a standard deviation of 1.0 to avoid unequal weighting due to different ranges of variation. Factor scores were calculated from eqn (1). Between-station/within-station analysis of variance is summarized also in Table 6. All factors rejected the null hypothesis at or below the 0.01 significance level.

3.2.2. *Naquadat Factor Results*

Nineteen physical-chemical parameters were measured by the Naquadat monitor and laboratory chemical analyses. The primary 16 station × 19 parameter data matrix was produced. The 19 × 19 character intercorrelation matrix was calculated and analysed by the factor analytic methods discussed previously. Four extracted factors accounted for 90% of the total character variance. These four factors were rotated by varimax methods

5. ANALTYICAL METHOD IN MICROBIOLOGY   153

and are itemized in Table 4. Station factor scores are shown in Table 5. Within-station variance information was not available in the Naquadat data therefore the analysis of variance was limited to the between-month/within-month variance ratio summarized in Table 5. Again the four Naquadat factors rejected the null hypothesis at or below the 0.01 significance level.

Table 4. Physical and chemical parameters (Naquadat factors).

Factor no. and eigenvalue	Parameters	Factor loading	Interpretation
NQ 1 9.99	Specific conductance HCO_3^-, Cl^-, Na^+, K^+, Ca^{2+}, Mg^{2+}, SO_4^{2-}, NH_4^+	0.96 0.90–0.99 0.81–0.89	Specific conductance or ion concentration
NQ 2 4.35	Water temperature Carbon Phosphorus NO_2–NO_3	0.96 0.78 0.97 −0.74[a]	Temperature related nutrients
NQ 3 4.35	Dissolved oxygen (mg l^{-1}) percentage O_2 saturation pH	0.76 0.89 0.72	Oxygen concentration
NQ 4 0.60	Carbon dioxide	0.79	CO_2 concentration

[a] A negative loading.

3.2.3. Geometric Representation

The factors generated by these analyses represent the significant orthogonal dimensions of our study. Microbiological factors (MX) 1, 2, and 5, were chosen as the most biologically relevant dimensions of the microbiological factors, accounting for 65% of the common variance. These three orthogonal dimensions were used as axes in a three-dimensional space and the station factor scores projected onto this space. This plot is shown in Fig. 2 and condenses the information of Table 6. Similarly, Naquadat factor (NQ) 1, 2 and 3, accounting for 93% of the common variance condense the scores of Table 5 into Fig. 3.

Table 5. Station scores on Naquadat factors.

		Specific conductance	Temp. C, P	O_2	CO_2
		Factors: NQ 1	NQ 2	NQ 3	NQ 4
Sept.	12.00	1.200	−0.476	0.014	−0.766
	18.00	1.364	−0.653	0.203	−0.879
	24.00	1.294	−0.265	0.169	−0.777
	06.00	1.265	−0.376	0.239	−0.683
Oct.	12.00	0.427	−0.336	1.459	−0.568
	18.00	0.482	−0.073	1.108	−0.563
	24.00	0.545	−0.017	1.485	−0.588
	06.00	0.709	−0.460	0.361	−0.691
March	12.00	−0.473	−1.113	−0.110	−0.378
	18.00	−0.769	−0.893	−0.186	−0.039
	24.00	−0.626	−1.057	−0.308	−0.164
	06.00	−0.817	−0.633	−0.296	0.119
July	12.00	−1.225	1.725	−1.640	2.408
	18.00	−1.244	1.435	−1.016	1.014
	24.00	−0.973	1.345	−0.783	0.634
	06.00	−1.158	1.846	−1.701	1.922
Mean		0.000	0.000	0.000	0.000
S.D.		1.000	1.000	1.000	1.000
F-ratio[a]		333	110	74	24
Critical variation		0.3	0.5	0.6	1.0

[a]Within-month/between-month analysis of variance.

4. Interpretation of Factors

4.1. Microbiological Factors

The biological factors are described in Table 2. The interpretation of the factors is directly related to the individual test of loadings. In the interpretation of biological factor MX 1, for instance, 72 substrate utilization tests had loadings of 0.5 or greater. High negative loadings were observed for glucose fermentation, H_2S production and nitrate reduction. On the same

155

Table 6. Station scores on Meduxnekeag factors.

Month	Time	MX 1	MX 2	MX 3	MX 4	MX 5	MX 6	MX 7	MX 8
September	12.00	−0.196	1.101	−0.394	1.009	−0.896	0.802	0.189	0.281
	18.00	−0.310	1.108	−0.370	0.539	−0.536	2.165	0.354	0.475
	24.00	−0.372	1.143	−1.199	0.483	−1.004	0.557	−0.390	0.345
	06.00	−0.602	0.787	−0.125	−0.802	0.326	−0.360	−1.143	−0.081
October	12.00	2.644	0.885	1.015	−0.845	−1.702	0.962	1.224	−1.845
	18.00	2.360	1.182	1.586	−0.504	−1.109	0.074	1.442	2.807
	24.00	−0.596	−0.321	1.148	−1.945	1.429	−1.553	−1.083	−0.958
	06.00	0.279	1.237	2.379	2.761	−0.634	1.557	0.997	1.195
March	12.00	−0.174	−1.059	0.013	0.046	−0.401	−0.558	−1.444	0.055
	18.00	−0.443	−1.455	−0.505	−0.259	−0.208	−0.318	−0.294	−0.541
	24.00	−0.362	−0.972	−0.379	−0.055	−0.117	−0.703	0.727	−0.473
	06.00	−0.438	−0.676	−0.211	−0.117	−0.036	0.093	1.372	−0.047
July	12.00	−0.496	−0.994	−0.944	−0.249	1.304	−0.755	−1.523	−0.368
	18.00	−0.539	−0.857	−0.560	−0.377	1.305	−0.221	−0.776	−0.176
	24.00	−0.352	−0.766	−0.608	−0.067	1.059	−0.418	−0.020	−0.215
	06.00	−0.403	−0.253	−0.845	0.383	1.228	−1.324	0.367	−0.455
Mean		0.000	0.000	0.000	0.000	0.000	0.000	0.000	0.000
S.D.		1.000	1.000	1.000	1.000	1.000	1.000	1.000	1.000
F-ratio[a]		51.7	25.1	22.2	12.0	27.8	16.0	11.2	8.8
Critical variation		0.4	0.6	0.6	0.8	0.6	0.7	0.8	0.9

[a] Within-station/between-station analysis of variance.

factor, positive loadings were observed for oxidase. Factor MX 1 has been interpreted as oxidative metabolism linked with nutritional versatility.

Within factor 1 the feature, glucose oxidation, demonstrated a negative loading at the 0.5 level whereas the loading for fermentation was −0.9 giving more weight to this variable. The overwhelming variables loading on this factor which are associated with aerobes supported the oxidative interpretation.

Factor MX 2 was interpreted as salt tolerance associated with oxidative metabolism.

The third MX factor was interpreted overall as lipid catabolism. Dye tolerance and growth at pH 4.0 were also included. This factor also carried an interesting negative loading for growth at 43°C, indicating lack of ability to grow at this temperature. The cold temperature factor, MX 4 appeared also to be linked with polymer production. The loading of 2-ketogluconate on this cold-water factor was quite fascinating as Lynch *et al.* (1975) has shown that 2-ketogluconate accumulated at 5°C in preference to other intermediates in the metabolism of glucose. At temperatures of 20°C, 25% of the glucose metabolized appeared as 2-ketogluconate in cultures of *Pseudomonas fluorescens*. It is possible that 2-ketogluconate may accumulate in these cold waters thus forming a natural substrate.

Nitrate reduction, the interpretation of factor 5, and nitrification, the interpretation of factor 6, did not have sufficient numbers of support variables to make the interpretation secure at this stage of the analysis. Nevertheless examination of the factor scores where the interrelationships between the variables are shown indicated that the interpretation was valid, i.e. a high score appeared in the station where it would be expected.

Factor 7, growth on lower nutrient levels, was too weak a factor for meaningful interpretation but did display some relevant correlations at a later stage. The loadings on factor 8 indicated that oxidative metabolism related to fluorescence would be an appropriate interpretation.

4.2. Physico-chemical Factors (Naquadat) (Table 4)

The factor accounting for most of the variance was the specific-conductance/ion-concentration factor. The factor NQ 2 was clearly water temperature and those nutrients, carbon and phosphorus which are most often related to temperature. The third factor, oxygen concentration was accompanied by pH. In these waters, pH varies directly with O_2 concentration and the factor analysis demonstrated this feature. Factor NQ 4, CO_2, was not highly significant in this phase of the analysis, however, due to the biological importance of CO_2 as an indirect measure of algal activity, the factor was retained.

4.3. Interpretation of Microbial Factor Scores

The second phase of interpretation was the examination of the factor scores as they related to each month, season and time of day. Factor 1, nutritional versatility/oxidative metabolism, scored negatively in the moderate range for all stations except three in October (Table 6).

Very little diurnal variation was observed in March and July and the variation was only slightly increased in September. October scores indicated diurnal fluctuation in nutritional versatility and oxidative metabolism. This diurnal variation is clearly demonstrated in Fig. 2. The October 12.00 and 18.00-h stations are at maximum and almost maximum projection; the spire representing October 06.00 h is just barely over the zero line and remaining stations project just under the zero line. This factor remained stable for most of the samples tested.

Much more variation was observed in salt tolerance. All of the September samples and three of the October samples projected high on this factor. The lowest projections were observed in March.

Factor MX 5 was projected into the hyperspace and the relative positions of the stations are indicated by spire height. The July spires are all well above zero, however, the 24.00-h October sample obtained the maximum value observed. The scores of each station in each month can be utilized to confirm the initial interpretation of the factors. The scores are also utilized to relate the factors to other factors or to untreated data. That is data which has not been factor analysed but which is related to the analysed database.

Returning to Table 6, where each factor score has been recorded, it appears that factor MX 3, lipid catabolism, demonstrated a definite diurnal variation in both September and October. The catabolism of lipids is a distinctive characteristic within the bacteria and are helpful substrates in taxonomic schemes. The scores for the March populations are negligible and would have been so in September also if it had not been for the high negative score at 24.00 h.

Factor MX 4, cold temperature/polymer production, scored a high positive at 12.00 h in September and at 06.00 h in October and a high negative at 24.00 h in October. The expected high positive scores for March did not materialize, in part due to the numbers of mesophiles in cold water (Bell et al., 1980), and in part due to the emphasis on cold temperatures in the interpretation. The interpretation will not be complete on several of these factors until substantiated by showing the relationship related to other data sets either by mathematical correlation or by a more detailed examination of the biological data.

Factor MX 5, nitrate reduction, scored a high positive throughout July as expected, yielded a peak in October and low scores in March. Diurnal variation was observed in September and October.

Nitrification scores were highest in September at 18.00 h and October at 06.00 h. The score for this factor, MX 6, in October at 24.00 h was the reverse, a high negative.

Factor MX 7 varied seasonally and diurnally showing high negatives and positives in the same 24.00-h period without any seasonal trend. The oxidative/fluorescence factor, MX 8, was an important attribute in October demonstrating high negative and positive scores. The diurnal fluctuation of this factor did not parallel exactly the other oxidative factors 1 and 2. A closer look at the October scores indicates that the influence of the fluorescence variable was more important at 12.00 h, where MX 8 showed a high negative score and MX 1 a high positive.

4.4. Physico-chemical Naquadat (NQ) Factor Scores

The first factor scored high in September as shown in Table 5 and Fig. 2. Very little diurnal variation was observed. The October samples yielded positive but moderate scores and the seasonal fluctuations were apparent. March and July produced negative scores throughout. Factor NQ 2, water temperature/nutrients, yielded high positive scores in July and high negative scores in March with negligible negative scores in October and September when the water temperature was 10–11°C.

The oxygen/pH factor scores were high and positive in October and high and negative in July. The scores on this factor are predictable in September, October and July from the untreated data. However, March scores are negligible. The spire height of March stations is low, not as low as July but similar to September. A careful re-examination of the raw data provided the explanation that the influence of the pH variable was greater than expected. Oxygen was such a powerful environmental parameter in this river that the pH which parallels O_2 is often neglected.

Factor NQ 4, CO_2, was high in July and negative or negligible for most of the other seasons. This is predictable also.

The score data then is acceptable as far as it goes, i.e. it does not violate any known principles. On the one hand it assists in the interpretation of the factors, tempering the import placed on one variable and stressing another in the interpretation of the hypothetical constructs. The seasonal and diurnal changes are amplified and we have a set of scores which enable us to compare these variables with the biological variables.

In summary, the most significant seasonal influences according to the scores are specific conductance in September, oxygen/pH in October, high temperature and the nutrients, carbon and phosphorus, and CO_2 in July and low temperature and low nutrients in March. In July there is a strong suggestion of free or aggressive CO_2, biodegradation and algal respiration.

4.5. Correlation of Physico-chemical with Biological Factors

Using Pearson product moment correlation as shown in Table 6 the complete array of correlations are presented. The significant correlations have been selected and displayed in Table 7 to assist in the interpretation as shown in Table 8.

Several features of the environmental influence are predictable; these are primarily the correlation of the oxygen concentration with oxidative metabolism and nutritional versatility; and secondly the strong linkage between microbial nitrate reduction and the temperature, nutrients and CO_2 of the water. Nitrate reduction is clearly influenced by O_2 concentration. The correlation between CO_2 and nitrate indicated a lowering of oxygen levels in July.

Certain factors and correlations were unexpected in this freshwater river, particularly specific conductance and ion concentration as primary influences. The highest correlation was observed between the biological factor salt tolerance/oxidative metabolism and the physico-chemical factor specific conductance. The positive correlations with O_2 and negative correlations with CO_2 completed the picture.

Another feature of the population which might have gone unnoticed in another type of analytical procedure was the factor lipid catabolism and the dependence of this ability on O_2. Very little is known of the role of lipid catabolism in natural systems.

Nitrate reduction in receiving waters is linked to several aspects of the bacterial population. Nitrate acts as an alternate hydrogen acceptor which increases the flexibility of aerobic bacteria to survive the lowering of O_2 concentration. Nitrate also enters the food chain and is an important nutrient adjunct in this nitrogen-limited water. Nitrification, on the other hand, is under the influence of specific conductance as well as oxygen, an unexpected result of the analysis. Nitrification is an important contribution of the bacteria to the ecosystem and appears in this analysis by the presence of ammonium ion in factor NQ 1 as a significant influence. Certain ecologically important aspects of bacterial activity, notably nutritional versa-

Table 7. Pearson correlation coefficients.

Naquadat factors	\multicolumn{8}{c}{Biological factors}

Naquadat factors	MX 1	MX 2	MX 3	MX 4	MX 5	MX 6	MX 7	MX 8
NQ 1	0.2230 $p = 0.203$	0.8629 $p = 0.0000$	0.3365 $p = 0.101$	0.1728 $p = 0.261$	−0.5899 $p = 0.008$	0.5944 $p = 0.008$	0.1227 $p = 0.325$	0.2912 $p = 0.137$
NQ 2	−0.1445 $p = 0.297$	−0.2157 $p = 0.211$	−0.3077 $p = 0.123$	−0.1081 $p = 0.345$	0.6954 $p = 0.001$	−0.3897 $p = 0.068$	−0.2403 $p = 0.185$	−0.1127 $p = 0.339$
NQ 3	0.5582 $p = 0.012$	0.5886 $p = 0.008$	0.8114 $p = 0.000$	−0.0392 $p = 0.443$	−0.5951 $p = 0.008$	0.4070 $p = 0.059$	0.3421 $p = 0.097$	0.1842 $p = 0.247$
NQ 4	−0.2798 $p = 0.147$	0.5788 $p = 0.009$	−0.4615 $p = 0.036$	−0.0934 $p = 0.365$	0.7046 $p = 0.001$	−0.5300 $p = 0.07$	−0.2721 $p = 0.154$	−0.2438 $p = 0.181$

Table 8. Correlations of bacterial factors with environmental factors.

Bacterial factor no.	Interpretation of bacterial factors	Pearson coefficient	Naquadat factor no.	Interpretation of environmental factors
MX 1	Oxidative metabolism Nutritional versatility	0.56 $p = 0.01$	NQ 3	Oxygen concentration
MX 2	Salt tolerance Oxidative metabolism	0.86 $p = 0.00$	NQ 1	Specific conductance ion concentration
		0.59 $p = 0.00$	NQ 3	Oxygen concentration
		−0.59 $p = 0.001$	NQ 4	CO_2
MX 3	Lipid catabolism	0.81 $p = 0.00$	NQ 3	Oxygen concentration
MX 5	Nitrate reduction	0.71 $p = 0.00$	NQ 4	CO_2 concentration
		0.70 $p = 0.00$	NQ 2	Temperature Carbon, Phosphorus
		−0.60 $p = 0.00$	NQ 3	Oxygen concentration
		−0.59 $p = 0.00$	NQ 1	Specific conductance
MX 6	Nitrification	0.60 $p = 0.01$	NQ 1	Specific conductance
		−0.53 $p = 0.02$	NQ 4	CO_2

Table 9. Predominant bacteria in order of cluster size.

Time	September	October	March	July
Noon	Aeromonas sp. (18)* Aeromonas hydrophila (11) Aeromonas hydrophilia hydrophila (10) Aeromonas hydrophila anaerogenes (3) Pseudomonas aeruginosa (3)	Pseudomonas caryophylli (74) Xanthomonas campestris (18)	Pseudomonas fluorescens I (6) Pseudomonas stuzeri (2) Pseudomonas lemoignei (2)	Flavobacterium lutescens (4) Flavobacterium aquatile (3) Pseudomonas aeruginosa (3) Flavobacterium capsulatum (2) Corynebacterium sp. (2) Plesiomonas shigelloides
Early evening	Pseudomonas alcaligenes (24) Aeromonas hydrophila anaerogenes (16) Aeromonas hydrophila proteolytica (11) Aeromonas hydrophila (2) Erwinia quercina (3)	Pseudomonas fluorescens II (29) Pseudomonas fluorescens IV (24) Pseudomonas caryophylli (2) Flavobacterium aquatile (2)	Pseudomonas sp. (2) Pseudomonas lemoignei (2) Corynebacterium (2)	Flavobacterium breve (4) Pseudomonas alcaligenes (3) Pseudomonas facilis (2)
Midnight	Aeromonas hydrophila anaerogenes (25) Aeromonas hydrophila proteolytica (18) Aeromonas hydrophila hydrophila (2)	Flavobacterium devorans (39) Flavobacterium rigense (13) Flavobacterium breve (3) Flavobacterium aquatile (3) Plesiomonas shigelloides (2)	Pseudomonas fluorescens III (2) Plesiomonas sp. (2)	Cytophaga johnsonae (3) Pseudomonas lemoignei (2) Pseudomonas stuzeri (2) Vibrio sp. (2) Alcaligenes sp. (2) Citrobacter intermedius (2) Gluconobacter sp. (2) Pseudomonas alcaligenes (2)

Early morning	*Flavobacterium rigense* (20) *Flavobacterium devorans* (10) *Plesiomonas shigelloides* (7) *Aeromonas hydrophila anaerogenes* (5) *Flavobacterium breve* (5) *Flavobacterium aquatile* (4) *Aeromonas hydrophila proteolytica* (3) *Pseudomonas aeruginosa* (2)	*Pseudomonas fluorescens* I (18) *Flavobacterium rigense* (9) *Pseudomonas lemoignei* (9) *Aeromonas hydrophila hydrophila* (6) *Aeromonas hydrophila anaerogenes* (5) *Pseudomonas cichorii* (4) *Pseudomonas fluorescens* III (3) *Flavobacterium lutescens* (2) *Pseudomonas stutzeri* (2) *Pseudomonas putida* (2) *Pseudomonas marginata* (2)	*Flavobacterium rigense* (8) *Pseudomonas pseudoalcaligenes* (3) *Pseudomonas acidovorans* (2) *Pseudomonas lemoignei* (2)	*Erwinia* sp. (13) *Pseudomonas* sp. (5) *Flexibacter flexilis* (3) *Flexibacter succinicans* (3) *Azotobacter chroococcum* (3) *Cytophaga fermentans* (3) *Aeromonas hydrophila anaerogenes* (2) *Xanthomonas campestris* (2) *Pseudomonas acidovorans* (2) *Flavobacterium rigense* (2) *Alcaligenes eutrophus* (2) *Pseudomonas solanacearum* (2)
Evenly distributed	*Aeromonas hydrophila anaerogenes* (3) *Plesiomonas shigelloides* (2) *Flavobacterium rigense* (2)	*Pseudomonas caryophylli* (4)	*Flavobacterium lutescens* (2) *Pseudomonas lemoignei* (2) *Alcaligenes* sp. (2)	*Flavobacterium breve* (9) *Aeromonas hydrophila shigelloides* (6) *Flavobacterium aquatile* (5) *Vibrio anguillarum* (3) *Plesiomonas* sp. (2) *Pseudomonas lemoignei* (2) *Aeromonas punctata punctata* (2) *Xanthomonas campestris* (2)

* (number of strains in cluster)

tility and biodegradation are assisted by oxygen and a high ion concentration.

A brief look at the correlations for factors MX 7 and MX 8 show that these less-important factors do not correlate well with any Naquadat factor. There is some indication that growth at lower nutrient levels is stimulated by oxygen which fits the nutritional versatility concept. Factor 8 appears to be a pale reflection of factor 1. A repeat study of this river has included more features related to low nutrients and nitrogen metabolism (Bell et al., 1980). The environmental parameters have been expanded and the algal population was represented by more variables.

4.6. Diurnal and Seasonal Changes in the Predominant Bacterial Species

The species shown in Table 9 are those which clustered using the similarity coefficient (S_J) of Sneath, UPGMA clustering and identification of the clustered strains by Bergey's manual (Buchanan and Gibbons, 1975) supported by a rapid DNA analysis method developed for these studies (Cashion et al., 1978). One hundred isolates from each sampling time (synonymous with station) were tested on 230 tests. The details of the numerical analysis have been reported by Cormier (1978) and Holder-Franklin (1981) previously.

The predominant species in September were *Aeromonas hydrophila* which subclustered according to subspecies. The *Flavobacterium* genus was also well represented in the predominant group. Although the early morning sample displayed a greater diversity of species, the majority of strains were facultative anaerobes. *Pseudomonas alcaligenes* although oxidase positive is in our experience commonly found with the fermenters, preferring higher temperatures. *P. aeruginosa* is often identified as a pollution indicator. The group selected in this econiche as predominant has the general characteristics of organisms which can tolerate higher salt concentrations, have an oxidative capability and do not thrive at 4°C. The temperature of the water samples was 10°–12°C. The factor scores of this group on factor 1 are negative indicating a low nutritional versatility which in fact agrees with the test findings. Therefore one concludes that the emphasis in the interpretation of MX 1 should be placed on nutritional versatility. The station scores for September on factor 2 were in the high positive range reflecting the salt tolerance of the strains. The drop in the station scores observed at 06.00 h does show the influence of the oxidative metabolism component of the factor as the most anaerobic group was isolated at 06.00 h.

Although the October samples were obtained only 2 weeks later and the water temperature was identical, a very different picture was observed in the species isolated.

The organisms were highly aerobic at noon and early evening but at midnight a shift to facultative anaerobes was seen. By early morning, more aerobic species were observed and the fluorescent *Pseudomonas* biotypes were dominating. The diurnal changes were also profiled by the factor scores. Factor MX 1 scores were highly positive at noon and early evening when the aerobic, nutritionally versatile organisms predominated, became moderately negative when the non-versatile facultatively anaerobic group predominated and shifted back to low positive when the mixture of aerobes and facultatives appeared.

October MX scores again reflected the diurnal changes. The mixture of oxidative metabolism and salt tolerance is much more complex. At 12.00 h the scores were moderately positive and the organisms clearly oxidative but not strongly salt tolerant. The increase in oxidative organisms in the early evening yielded a high score. The complete change in species at midnight to *Flavobacterium* brought the score to the negative side. By early morning, the mixture of aerobes, facultatives and salt-tolerant organisms was apparent in the high score on factor MX 2.

The interrelationships of oxygen and salt tolerance were difficult to follow in this part of the analysis.

The comparison of species to factors is a useful exercise when substantiation for factor interpretation is sought, however, this type of laborious and not always productive comparison can be replaced by a mathematical correlativity.

4.7. Correlations of Biological Factors with Environmental Factors

The factor scores generated by the microbiological data were correlated with the Naquadat factor scores.

Instead of comparing the factor scores with the raw environmental data we analysed this data for factors separately. In most factor analyses, the physico-chemical data is either included in the analysis with the biological data or is correlated directly without previous factor analysis. The most manageable, accurate and balanced method is to factor analyse the physico-chemical data separately and bring the two data sets together either by regression analysis or Pearson product moment correlation. This feature is one of the unique characteristics of our method.

The complete set of correlations is shown in Table 6 and the most significant correlations displayed in Table 7. Oxidative metabolism and

Table 10. Summary of conclusions from factor scores.

	Predominant genera	Highest biological factor scores	Highest Naquadat factor scores	Conclusions
September	*Aeromonas* *Flavobacterium*	Salt tolerance Oxidative metabolism (all times) Nitrification (at 18.00 h)	Specific conductance Oxygen concentration (all times)	Salt tolerance, oxidative metabolism and nitrification are related to the oxygen concentration of the water. Salt tolerance among the bacterial population related to the ion concentration of the water
October 12.00 h 18.00 h 24.00 h 06.00 h	*Pseudomonas carophylli* *Pseudomonas fluorescens* *Flavobacterium* Mixed flora;[a] *Pseudomonas*	Nutritional versatility Oxidative metabolism (12.00 h; 18.00 h) Salt tolerance Oxidative metabolism	Oxygen concentration Negative scores for CO_2 conc. Specific conductance (all times)	Diurnal variation and population changes apparent. Strong links seen between oxidative metabolism and oxygen concentration. Variations seen in salt tolerance
March	*Pseudomonas*	Negative scores for salt tolerance Oxidative metabolism	Low temperature Low carbon and phosphorus levels High nitrate	Very heterogeneous population; no diurnal change. Bacteria living in low-nutrient, low-temperature environment
July	*Flavobacterium* *Pseudomonas* *Cytophaga* *Aeromonas*	Nitrate reduction	Free or aggressive CO_2 Biodegradation Algal respiration	Diurnal variation was less than expected. Low oxygen levels were linked to nitrate reduction. Evidence of stable cycle due to algal activity observed

[a] Note Table 9.

nutritional versatility correlates with the oxygen concentration of the water. In other words, one of the most important features of the bacterial population changes in relation to the oxygen concentration of the water, the factor MX 2, salt tolerance and oxidative metabolism, has double links to specific conductance (ion concentration) and oxygen. It is not possible to determine which is the more important influence at this stage as both ion and oxygen concentrations seem heavily weighted and the biological characteristics absorbs 90% of the variance.

Lipid catabolism is highly correlated to oxygen concentration. There is also a temperature component to factor MX 3 with high positive loadings for 4°C and a high negative loading for 43°C. Looking back at Table 5, Meduxnekeag factor 3 has high scores for the October samples which were primarily *Pseudomonas* strains. The first three MX factors are linked to the *Pseudomonas* group through nutritional versatility, oxidative metabolism and lipid catabolism. MX 4 does not correlate with any environmental factor. MX 5, nitrate reduction, is profoundly influenced by all of the environmental factors. The reduction of nitrate increases with increase in CO_2, temperature and the nutrients, carbon and phosphorus and negatively with oxygen concentration and specific conductance. This strong correlation is related to the seasonal fluctuations in nitrate particularly to the build up of NO_3 in the winter. The prominence of nitrate in this analysis is a reflection of the overall nitrogen limitation in this river, the high concentrations of oxygen which inhibit nitrate reductase and the importance of nitrate as a source of nitrogen. This feature of the interrelationship between the biological and the environmental characteristics is revealed by the analysis.

The conclusions are summarized in Table 10 where the high seasonal and diurnal factor scores which demonstrated the greatest correlation to the high physico-chemical scores have been tabulated along with the predominant genera. The predominant genera in September are facultative and salt tolerant. In October, the diurnal fluctuations in bacteria change with changes in salt tolerance. The environmental changes must have been quite subtle in October as evidenced by the consistently high scores for oxygen and ion concentraton in all physico-chemical samples. The March samples clearly demonstrate that very few correlations can be made between bacterial species and environmental change when a small number of strains cluster. There is little to be gained from identifying every isolate if there is a very heterogeneous population. This heterogeneity may be an indicator of a very stable population.

In July the scores emphasize biodegradation and nitrate reduction with many implications of algal activity. These features are not seen in the colder samples, i.e. below 15°C.

5. Summary

A complete and detailed description of one possible type of factor analysis using research data obtained on bacteria isolated from a natural environment has been given. In addition, an outline of other possible approaches has been provided. The use of factor analysis to obtain a mathematical solution to an environmental problem has been described. The biological processes which appear to be operating in the freshwater econiche can be affirmed by the loadings of certain characteristics on the factors, followed by the revelation of interrelationships between the variables by the factor scores and finally by correlating the environmental and biological factors. The essential phase of the analysis for the full interpretation of both data sets, the physico-chemical and the biological, is the demonstration of a relationship between the bacterial population and the environment.

References

Bell, C. R., Holder-Franklin, M. A. and Franklin, M. (1980). *Water Res.* **14**, 449–460.
Bell, C. R., Holder-Franklin, M. A. and Franklin, M. (1982a). *Appl. Environ. Microbiol.* **43**, 269–283.
Bell, C. R., Holder-Franklin, M. A. and Franklin, M. (1982a). *Canad. J. Microbiol.* **28**, 959–975
Buchanan, R. E. and Gibbons, N. E. (1975). *In* "Bergey's Manual of Determinative Bacteriology." Williams and Wilkins, Baltimore.
Cashion, P., Holder-Franklin, M. A., McCully, J. and Franklin, M. (1977). *Anal. Biochem.* **81**, 461–466.
Comrey, A. L. (1973). *In* "A First Course in Factor Analysis." Academic Press, New York.
Cormier C. J. (1978). M.Sc. Thesis, University of New Brunswick, Fredericton, New Brunswick.
Harman, H. H. (1967). *In* "Modern Factor Analysis." University of Chicago Press, Chicago.
Holder-Franklin, M. A. (1981). *In* "Methods of Studying Population Shifts in Aquatic Bacteria in Response to Environmental Change." Scientific Series, No. 124, Inland Waters Directorate, Water Quality Branch. Department of the Environment, Ottawa, Canada.
Holder-Franklin, M. A., Franklin, M., Cashion, P., Cormier, C. and Wuest, L. (1978). *In* "Microbial Ecology" (M. W. Loutit and J. A. R. Miles, eds.), pp. 45–50. Springer-Verlag, New York.
Howarth, R. J. and Murray, J. W. (1969). *J. Paleontol.* **43**, 660–675.
Jaccard, P. (1908). *Bull. Soc. Vand. Sci.* **44**, 223–270.
Lynch, W. E., MacLeod, J. and Franklin, M. (1975). *Can. J. Microbiol.* **21**, 1560–1572.

Rohlf, F. J., Kishpaugh, J. and Kirk, D. (1971). Technical Report State University of New York at Stony Brook, New York.
Rosswall, T. and Kvillner, E. (1978). *In* "Advances in Microbial Ecology" (M. Alexander, ed.) Vol. 2, pp. 1–48. Plenum, New York and London.
Sneath, P. H. A. and Sokal, R. R. (1962). *Nature, Lond.* **193**, 855–860.
Sundman, V. (1970). *Can. J. Microbiol.* **16**, 455–464.
Sundman, V. (1973). *Bull. Ecol. Res. Commun. Stockholm* **17**, 135–141.
Sundman, V. and Gyllenberg, H. G. (1967). *Ann. Acad. Sci. Fenn.* Ser. A IV, **112**, 1–32.
Toerien, D. F. (1970) *Water Res.* **4**, 305–314.
Toerien, D. F., Hattingh, W. H. J., Kotze, J. P., Thiel, P. G. and Sibert, M. L. (1969). *Water Res.* **3**, 129–140.
Väätänen, P. (1980) *Appl. Environ. Microbiol.* **40**, 55–61.

Chapter 6

Process Analysis in Microbial Systems: Biofilms as a Case Study

W. G. Characklis

College of Engineering, Montana State University, Bozeman, Montana, U.S.A.

The first two-thirds of this chapter describe the principles of process analysis and in doing so serve to define and extend much of the material on modelling introduced in Chapter 2. In the last third of the chapter the author applies the methods of process analysis to bacteria growing on surfaces in biofilms, a topic of significant commercial importance—*Editorial note.*

1. Introduction

Bacteria stick firmly, and often with specificity, to almost any surface submerged in an aqueous environment. The bacterial attachment is mediated by polymeric material, primarily polysaccharide, that extends from the cell to form a tangled mass of fibres, termed a *glycocalyx* (Costerton *et al.*, 1978). The adhesion mediated by the glycocalyx may provide an attractive habitat for certain bacteria in many aquatic environments. The cells grow, reproduce and form products at the surface which increase the accumulation of the organic layer. The entire deposit is termed a *biofilm*.

Biofilm processes may be beneficial as exemplified by fixed-film wastewater treatment processes (e.g. trickling filters and rotating biological contactors). In addition, biofilms frequently play a major role in stream purification processes. In fact, microbial activity in natural waters has been found predominantly at interfaces (Marshall, 1976; Costerton *et al.*, 1978). However, biofilms can be quite troublesome in certain engineering systems. For example, biofilms in water conduits can cause energy losses resulting from increased fluid frictional resistance and increased heat transfer resistance. Numerous concerns regarding biofilms have stimulated further research and analysis of the processes (Table 1).

This chapter begins with an introduction to process analysis followed by its application to biofilm processes. The structure and properties of biofilms, especially those which contribute to stoichiometric or kinetic models, are

Table 1. Effect and relevance of biofilms on various rate processes.

Effects	Specific process and result	Concerns
Heat transfer reduction	Biofilm formation on condenser tubes and cooling tower fill material. *Energy losses*	Power industry Chemical process industry U.S. Navy Solar energy systems
Increase in fluid frictional resistance	Biofilm formation in water and wastewater conduits as well as condenser and heat exchange tubes. Causes increased power consumption for pumped systems or reduced capacity in gravity systems. *Energy losses*	Municipal utilities Power industry Chemical process industry Solar energy systems
	Biofilm formation on ship hulls causing increased fuel consumption. *Energy losses*	U.S. Navy Shipping industry
Mass transfer and chemical transformations	Accelerated corrosion due to processes in the lower layers of the biofilm. Results in *material deterioration* in metal condenser tubes, sewage conduits and cooling tower fill	Power industry U.S. Navy Municipal utilities Chemical process industry
	Biofilm formation on remote sensors, submarine periscopes, sight glasses etc. causing *reduced effectiveness*	U.S. Navy Water quality data collection

Detachment of micro-organisms from biofilms in cooling towers. Releases *pathogenic organisms* (e.g. *Legionella* in aerosols)	Public health
Biofilm formation and detachment in drinking water distribution systems. Changes *water quality* in distribution system	Municipal utilities Public health
Biofilm formation on teeth. Causes *dental plaque and caries*	Dental health
Attachment of microbial cells to animal tissue. Causes *disease* of lungs, intestinal tract and urinary tract	Human health
Extraction and oxidation of organic and inorganic compounds from water and wastewater (e.g. rotating biological contacters, biologically-aided carbon adsorption and benthal stream activity). *Reduced pollutant load*	Wastewater treatment Water treatment Stream analysis
Biofilm formation in industrial production processes *reduces product quality*	Pulp and paper industry
Immobilized organisms or community of organisms for conducting *specific chemical transformations*	Chemical process industry
Fouling biofilm accumulation *reduces effectiveness* of ion exchange and membrane processes used for high-quality water treatment	Desalination Industrial water treatment

presented. The various rate processes which contribute to biofilm development are described and mathematical expressions are presented within the context of mass balances. Finally, the effects of biofilms on energy losses are presented, including the effect on fluid frictional resistance (momentum balances) and the effect on heat transfer resistance (energy balances).

2. Process Analysis

Process analysis refers to the application of scientific methods to the recognition and definition of problems and the development of procedures for their solution. This generally requires (i) mathematical specification of the problem for the given physical situation, (ii) development of a mathematical model and (iii) synthesis and systematic presentation of results to ensure full understanding. The *process* denotes an actual series of operations or treatment of materials as contrasted with the *model* which is a mathematical description of the process (Himmelblau and Bischoff, 1968).

2.1. Modelling (after Himmelblau and Bischoff, 1968)

A model represents a part, frequently a very simplified part, of reality. The models formulated in this book will represent only those aspects of reality which are of interest to us. Manipulation of the models improves our insight regarding the workings of the real system. Models are used in attempts to describe and understand the behaviour of real systems. However, real systems, even simple ones, are very complex and rarely understood. Since engineers and scientists have to deal with real systems, describe real systems and explain or make use of even simple real systems, models are used.

Modelling is an iterative process. First, a model is formulated in mathematical terms. The first model is usually too simple to be realistic. However, the model serves as a hypothesis and a set of experiments is designed to test the hypothesis. If the results of experiments differ significantly from predictions, the first model is modified and a new hypothesis must be formulated.

A general procedure for the analysis of complex processes has been suggested by Himmelblau and Bischoff (1968):

(1) Formulation of the problem(s) and establishment of objectives and criteria of value; delineation of performance requirements.

6. PROCESS ANALYSIS IN MICROBIAL SYSTEMS

(2) Preliminary inspection and classification of the process to break it down into subsystems (elements).
(3) Preliminary determination of the relationships among the subsystems.
(4) Analysis of the variables and relationships to provide as simple and consistent a set as possible.
(5) Mathematical modelling (in applicable cases) of the relationships in terms of the variables and parameters; description of elements that can only be incompletely represented by mathematical models.
(6) Evaluation of how well the model represents the real process, using judgement to integrate the non-mathematical with the mathematical representations.
(7) Application of the model; interpretation and comprehension of the results.

These steps are designed to develop an approach of structuring and analysing processes wherever possible through mathematical models. This

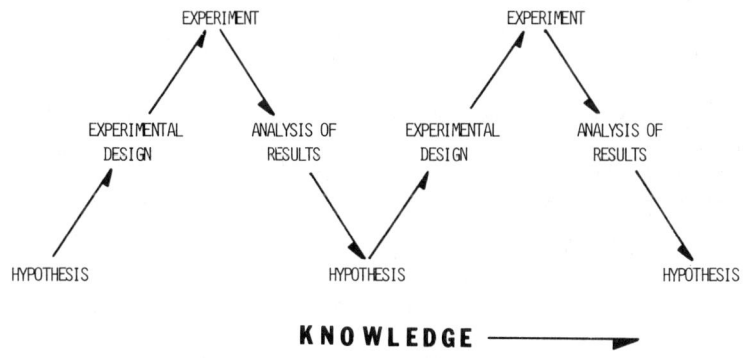

Fig. 1. The cyclical nature of model building.

approach provides for more rigorous analysis and tends to make subjective judgements (when required) more formal and thorough. Figure 1 indicates the cyclical nature of these steps.

A model never perfectly describes reality nor is that a necessity. However, the model must describe enough of reality to answer the questions that have been posed.

2.1.1. Types of Models

Three types of models and their combinations can be prepared:
(1) Transport phenomena models.
(2) Population balance models.
(3) Empirical models.

Examples of transport models are the phenomenological equations of change, that is, the equations describing the conservation of mass, momentum and energy. Residence time distributions and other age distributions are examples of population balance models. Finally, examples of typical empirical models are those polynomials used to fit empirical data by the method of "least squares". Discussions in this text will focus primarily on transport phenomena models and, to a limited extent, on population balance and empirical models.

Table 2 lists the classes of models along with a brief description and reference to their application in microbial systems. More detailed discussions can be found elsewhere (Roels and Kossen, 1978).

Table 2. Description of different types of mathematical model.

Class of model	Description and application
Descriptive (black box)	Curve fitting Interpolation of data Useful within the region where the model was tested experimentally *Example*: use of the logistic equation to describe batch microbial growth
Predictive (grey box)	Extrapolation of data With care, can be used outside the region tested experimentally *Example*: Michaelis–Menten (1913) equation for describing enzyme-substrate reaction rates
Structured	Systems in a transient state Useful for modelling changes in internal structure of cells or structure of mixed microbial populations *Example*: Williams (1967) two-compartment (synthetic component and genetic component) model for biomass

6. PROCESS ANALYSIS IN MICROBIAL SYSTEMS

Table 2. Description of different types of mathematical model.—*continued*

Class of model	Description and application
Unstructured	Does not account for internal structure Steady-state systems Lumped parameter models *Example*: Steady-state continuous flow microbial reactor (chemostat) where mass is the only property of the biotic material
Non-segregated	Considers that the biomass is homogeneously dispersed throughout the reactor fluid Mass concentration is the parameter which describes biomass Population balance models *Example*: Chemical oxidation of microbial cells by chlorine where biomass destruction or solubilization, not disinfection, is of greatest concern
Segregated	Uses a discrete parameter, the number of cells, as the parameter to describe biomass *Example*: Evaluating disinfection or sterilization processes where number of organisms alive and dead is most critical
Deterministic	Output variables have values completely determined by the structure of the system Applicable in cases where large numbers of organisms are involved ($n > 10\,000$) *Example*: activated sludge process for a chemostat in which saturation kinetics describes substrate removal and, consequently, microbial growth
Stochastic	For small numbers of organisms ($n > 100$) For cases where differences in organisms are important *Example*: different microbial species exhibit varying resistance to a sterilization process
Continuous time	Systems where events are evenly distributed in time Most microbial models *Example*: logarithmic growth rate model for a batch culture in which some micro-organisms are always undergoing division
Discrete time	Systems where events occur at a limited number of times *Example*: a synchronous culture in which all micro-organisms divide simultaneously at discrete times

2.2. Fundamentals Regarding Rate (after Churchill, 1974)

The fundamental relationships which underlie transport phenomena models are the equations for conservation of mass, momentum and energy which are a result of the laws of thermodynamics and Newton's laws of motion. The conservation equations are generally expressed in terms of intensive factors (i.e. independent of system mass) such as composition, velocity and temperature. The conservation equations also introduce quantities such as thermodynamic properties, stoichiometric coefficients, transport rates and chemical reaction rates (Table 3).

The thermodynamic properties, such as density, heat capacity and chemical equilibrium constants can be estimated with reasonable confidence from mechanistic models and can be measured with reasonable accuracy. The transport coefficients, such as viscosity, thermal conductivity and diffusivity, and the chemical reaction rate coefficients can rarely be predicted and are difficult to measure. Even the definitions of these latter quantities are somewhat arbitrary.

The equations for conservation of mass, energy, momentum and atomic species equate rate of accumulation to net rate of input by transport and net rate of input by various rate processes such as chemical reaction, diffusion, radiation, convection and viscous dissipation. Conservation equations or balance equations can be expressed in words as follows:

$$\begin{pmatrix} \text{net rate of} \\ \text{accumulation} \\ \text{in system} \\ \text{volume} \end{pmatrix} = \begin{pmatrix} \text{rate of} \\ \text{transport in} \\ \text{through} \\ \text{system} \\ \text{surface} \end{pmatrix} - \begin{pmatrix} \text{rate of} \\ \text{transport out} \\ \text{through} \\ \text{system} \\ \text{surface} \end{pmatrix} + \begin{pmatrix} \text{rate of} \\ \text{generation in} \\ \text{system} \\ \text{volume} \end{pmatrix} - \begin{pmatrix} \text{rate of} \\ \text{consumption} \\ \text{in system} \\ \text{volume} \end{pmatrix} \quad (1)$$

The change of some extensive parameter, E (an extensive quantity is proportional to the mass of the system), with time can be caused by two mechanisms:

(1) E can be generated or consumed within the system by some process, e.g. chemical reaction occurring at a rate per unit volume, r_E (r_E is negative if E is consumed).
(2) E can be transported into or out of the system through its boundary with the external environment.

2.2.1. Process Time and the Importance of Rate

The physical, chemical, and biological transformations which are discussed in this book are completed in a certain period of time. For example, the removal of soluble organics in a biological wastewater treatment process

Table 3. Intensive factors, physical properties and their relationships to the conservation equations.

	Conservation equations			
	Mass	Energy	Momentum	
Intensive variables	Concentration	Temperature	Velocity	Easy to measure accurate, precise
Physical properties				
Thermodynamic	Equilibrium constants	Heat capacity	Density	
Transport or reaction coefficients	Diffusivities and reaction rate	Thermal conductivity	Viscosity	Difficult to measure and define; not accurate or precise

occurs in a specified period of time; less than the hydraulic residence time within the reactor. A fouling biofilm accumulates on a heat exchanger surface over a period of time reducing heat transfer until a critical point is reached when the process unit must be shut down for cleaning operations. The time required for these specified changes to occur is inversely proportional to the rate at which the process occurs and is the most important quantity in process analysis.

2.2.2. The Anatomy of a Balance Equation

We will now reconsider the general balance equation presented in eqn (1) which can be reworded as follows:

$$\begin{matrix} \text{rate of} \\ \text{accumulation} \end{matrix} = \begin{matrix} \text{net rate of input} \\ \text{by transport} \end{matrix} + \begin{matrix} \text{rate of} \\ \text{process} \end{matrix} \qquad (2)$$

The process rates are fundamental quantities in that they can be generalized and correlated simply with factors such as temperature, pressure, composition, velocity and diameter which describe the environment. Process rate includes chemical reaction rate, adsorption rate, diffusion rate, rate of heat radiation or rate of viscous dissipation in fluids.

Rate of accumulation and *net rate of input by flow* are rates of change. These rates of change are easily measured and observed quantities which may be the result of several process rates. They cannot be easily correlated or generalized with factors that describe the environment.

It is essential that *rates of change* should not be confused with *process rates*. Rates of change are generally *measured* or *observed* quantities in a system from which process rates are *inferred*.

2.2.3. Mass Balance Equations

Mass balance equations will be the primary tool used to analyse microbial processes. Consequently, bulk transport rate, mass transfer rate and conversion rate will be encountered frequently. These rates can be classified as follows:

        transport or transfer rates,
          bulk transfer,
          mass transfer,
        conversion or transformation process,
          chemical conversions.

Bulk transport or *transfer* refers to the movement of material from one location to another as a result of flows. For example, bulk transport of oil from the well to the refinery through a pipeline.

6. PROCESS ANALYSIS IN MICROBIAL SYSTEMS

Chemical conversion or *transformation processes* refer to changes in composition as a result of chemical reactions. For example, the change in dissolved oxygen concentration in a fermenter due to respiration of bacterial cells.

Mass transfer describes the rate of material transport through a system boundary. The term mass transfer is usually reserved for interphase (e.g. solid–liquid or gas–liquid) transport processes. *Diffusion*, similar to mass transfer, generally describes intraphase transport. Consequently, the system and its boundaries must be specifically defined to analyse these processes.

One of the important criteria in defining transfer processes is that the molecular structure of the material remains unaltered. Transport processes are described by rather simple phenomenological equations containing either a concentration or concentration difference (for transport through the system boundaries) or concentration gradients (for transport within the system).

Transformation processes result in changes in molecular structure. For example, the metabolism of organisms results in the conversion of nutrients into cellular components, metabolites and heat. Transformation processes are described by *rate equations* or, more specifically, constitutive or kinetic equations of the following general form:

$$r = r(c_1, c_2, \ldots, c_n) \tag{3}$$

where c_i are concentrations of the various reacting components. Transfer and transformation processes will be combined in a general model based on the conservation equations as follows:

$$\underset{\text{transfer processes}}{\text{IN} - \text{OUT}} + \underset{\text{transformation processes}}{\text{CONVERSION}} = \text{ACCUMULATION} \tag{4}$$

2.3. Reactors (after Frederickson *et al.*, 1970)

Many biochemical processes involve *batch* growth of micro-organisms. After seeding a liquid medium of appropriate composition with an inoculum of living cells, nothing (except possibly some gas) is added to the culture or removed from it as growth proceeds.

However, many microbial processes of interest occur in *continuous flow* reactors which are characterized by continuous flow of reactants into the reactor while products are continuously removed. Two types of ideal continuous flow reactors will be important in our further discussions:

continuous flow stirred tank reactor (CFSTR),
plug flow reactor (PFR).

2.3.1. Residence Time Distribution of Fluid in Vessels

How can we describe the flow characteristics of fluid in a vessel adequately enough to yield information useful in the design or analysis of reactors? The approach is to find out how long individual molecules stay in the vessel by using population balance models. Information on the distribution of ages of molecules or particles in the exit stream or the distribution of residence times of molecules within the vessel can be found easily and directly by a widely used experimental technique, the stimulus-response technique. This information can then be used to account for the flow behaviour of fluid in a chemical flow reactor (Levenspiel, 1972).

In developing the "language" for the treatment of flow, consider the steady-state flow, without reaction and without density change (which includes almost all liquid phase systems), of a single fluid through a vessel. Under these conditions,

$$\theta = \frac{V}{F} = \text{holding time, mean residence time or space time} \quad (5)$$

$$D = \frac{F}{V} = \text{dilution rate or space velocity} \quad (6)$$

where V = reactor volume (L^3)
F = volumetric flow rate ($L^3 t^{-1}$)

For liquid phase continuous flow systems at steady state, the mean residence time of any system component, i, can be defined as follows:

θ_i = mean residence time

$$= \frac{\text{amount of component } i \text{ in the system}}{\text{mass flow rate of component } i \text{ out of the system}} \quad (7)$$

2.3.2. Continuous Flow Stirred Tank Reactor (CFSTR)

A continuous flow stirred tank reactor consists of a well-stirred tank into which there is a continuous flow of reacting material, and from which the (partially) reacted material passes continuously. The important characteristic of the CFSTR is the stirring.

In this type of reactor, culture medium and/or organisms are continuously

6. PROCESS ANALYSIS IN MICROBIAL SYSTEMS

fed to the reactor, where growth and other processes occur. The culture within the reactor is agitated somehow; generally, a mechanical agitator is provided for this purpose. Culture is removed from the vessel at the same volumetric rate as feed is admitted, so that culture volume remains constant.

In the ideal CFSTR, agitation is assumed to be so vigorous that mixing is complete. The criteria of complete or perfect mixing may be stated in different, but equivalent ways. The usual criterion is simply that the composition of the culture in the vessel is uniform, so that the composition of the stream leaving the fermenter is the same as that of any sample from the interior of the fermenter. From a statistical viewpoint, mixing is perfect:

(1) The probability that a "particle" (organism, molecule) will be in a given subvolume in the culture is the same as the probability that it will be in any other subvolume of equal size regardless of location in the vessel.
(2) "Particles" move independently through the vessel. The statistical criteria are the more general, and the first criterion given can be derived from them.

No real vessel can be perfectly mixed but, in practice, the behaviour of a real vessel can be made to approach that of an ideal CFSTR very closely.

Let c be the concentration (amount per unit volume of culture) of some substance, biotic or abiotic, in a culture. Then, application of the principle of conservation of mass to the system composed of the culture (liquid plus organisms) in the CFSTR yields the following differential equation:

$$V(\mathrm{d}c/\mathrm{d}t) = F(c_i - c) + rV \tag{8}$$

RATE PROCESS ANALYSIS

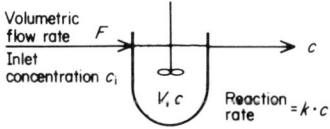

MATERIAL BALANCE ON COMPONENT c

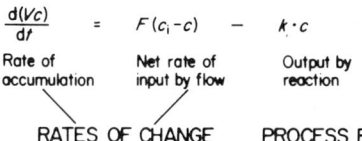

Fig. 2. Rate process analysis using material balances in a CFSTR.

relating the rate of change of c to the rate of flow through the vessel (F) and the rate of production of the substance per unit volume (r). In the equation, V is the volume of culture in the vessel (a constant), and c_i is the concentration of substance in the feed to the CFSTR (Fig. 2). The units of each term in the equation are mass per unit time (Mt^{-1}).

From this equation, follows the first point of importance concerning the CFSTR. Suppose that a steady state ($dc/dt = 0$) has been attained and that a substance is consumed or destroyed by reactions within the vessel ($r \neq 0$). Then we see that

$$D(c_i - c) = -r \qquad (9)$$

where $D = F/V$ is the *dilution rate*. This equation shows that the composition of the feed stream undergoes a discontinuous change when the stream enters the CFSTR. Thus, if the feed of the CFSTR contains organisms, they will experience a discontinuous change, or an environmental shock, upon entering the reactor. The equation also indicates that r can be determined from D, c_i and c which are measurable quantities.

A second point of importance regarding CFSTRs is that the residence time of a particle is not fixed but is subject to statistical fluctuation. The density distribution of residence time is

$$D\,e^{-Dt} \qquad (10)$$

The most probable residence time is zero. The mean residence time is D^{-1}. D^{-1} is also the standard deviation of residence time. If a cascade (series) arrangement of equal volume CFSTRs is used (Fig. 3), say m of

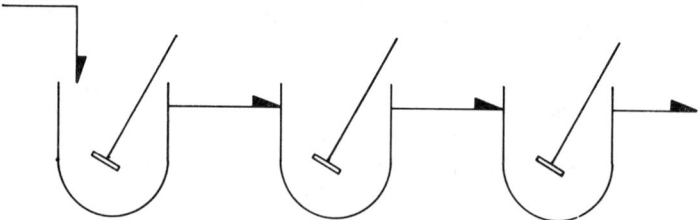

Fig. 3. A cascade (series) arrangement of equal volume CFSTRs.

them, then the residence time follows a gamma distribution with density function

$$(mD) \cdot [(mDT)^{m-1} \exp(-mDt)]/(m-1) \qquad (11)$$

The mean is again D^{-1} (V is the *total* volume of all m tanks) but the standard deviation is smaller than in the one-vessel case, and is $D^{-1}m^{-1/2}$.

6. PROCESS ANALYSIS IN MICROBIAL SYSTEMS

As the number of tanks in the cascade becomes larger and larger, while D is held constant (thus implying that the individual tanks become smaller and smaller), the standard deviation becomes smaller and smaller. In the limit of very large m, the residence time is no longer subject to statistical fluctuations, and is always D^{-1}. This arrangement of reactors is quite effective for simulating a plug flow reactor of volume V (see below).

Another point of importance concerning the CFSTR is that the probability that a particle in the vessel at some time will be "washed out" in some subsequent time interval is independent of such factors as the size of the particle or its residence time in the CFSTR. Since washout is equivalent to death so far as the population in the CFSTR is concerned, in effect, the flow through the vessel imposes a *non-selective* death rate on the population. This point becomes most important when two or more populations are growing symbiotically in a CFSTR.

2.3.3. Plug Flow Reactor (PFR)

In this type of reactor, culture medium and/or organisms are continuously fed to the reactor, where growth and other processes occur. The culture within the reactor is not stirred, since the object is to have elements of culture move progressively through the fermenter without mixing in the longitudinal direction.

In the ideal plug-flow reactor, adjacent elements of culture are assumed to move progressively through the tube without exchange of material between such elements. In addition, the composition of the culture is assumed to be uniform over any cross-section, though the composition obviously changes with distance through the reactor.

No real reactor can be an ideal PFR, but in some cases, the behaviour of real apparatus can be made to approach the ideal. A series of CFSTRs is quite effective as described above.

As before, let c be the concentration of some substance in the culture. Application of the principle of conservation of mass to a system of infinitesimal length moving with the velocity of flow through the tube (v) then yields the partial differential equation

$$(\partial c/\partial t) + v(\partial c/\partial z) = r \qquad (12)$$

where z is the axial distance from the inlet of the PFR, and r is the production rate per unit volume (Fig. 4).

In this case, the residence time of a particle is not subject to statistical fluctuations; a particle at position z has been in the vessel for a time z/v, and if the reactor has length L, the transit time will be L/v.

In many cases, mixing in the direction of flow may be important. This

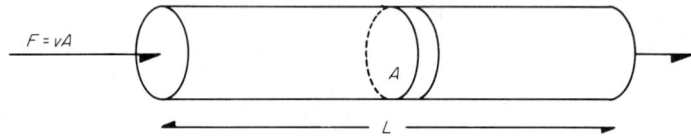

Fig. 4. A plug flow reactor.

is accounted for most easily (though only approximately) by introducing an effective axial diffusion coefficient, $\mathcal{D}$. The material balance (in this case, a differential material balance) leads to a more complicated equation:

$$(\partial c/\partial t) + v(\partial c/\partial z) = \mathcal{D}(\partial^2 c/\partial z^2) + r \qquad (13)$$

where the mixing term involves the second derivative of the concentration. The diffusivity $\mathcal{D}$ is not the usual diffusion coefficient that is used in studies of transport phenomena. $\mathcal{D}$ is dependent on the flow regime in the reactor and independent of the usual (molecular) diffusion coefficient, at least if flow is turbulent.

If axial mixing does occur, then particle residence times are again subject to statistical fluctuations. No simple expression for the density of the distribution of residence times emerges in this case [though it can be found by solving eqn (13) subject to appropriate boundary and initial conditions]. However, one can say that the mean residence time will be L/v and the standard deviation of the residence time will increase as $\mathcal{D}$ increases; the distribution will be skewed, with the most probable residence time being smaller than the mean. If the dimensionless parameter $\mathcal{D}/vL$ is small compared to unity, then the distribution of residence times is approximately Gaussian, with mean L/v and standard deviation $(L/v)(2\mathcal{D}/vL)^{1/2}$.

Biologically speaking, the most important feature of the PFR is the progressive *change* in environmental conditions seen by an organism traversing the reactor. This is in marked contrast to the constant conditions seen by an organism traversing an ideal CFSTR. In fact, the situation in the ideal PFR is entirely similar to that in the batch fermenter, with residence time, z/v, replacing batch reaction time, t. Hence, the ideal PFR can presumably accomplish on a continuous, steady-state basis that which is done discontinuously in batch reactors.

2.3.4. Segregation or Micromixing Effects

The problem associated with the mixing of fluids is important for extremely fast reactions in homogeneous systems, as well as for all heterogeneous systems. The major concern is the *degree of segregation* of the fluid, or

6. PROCESS ANALYSIS IN MICROBIAL SYSTEMS

whether mixing occurs on the microscopic level (mixing of the individual molecules) or the macroscopic level (mixing of clumps or aggregates of molecules). The degree of segregation does not affect batch reactors or plug flow reactors. However, it can significantly affect the performance of a CFSTR. The previously described stimulus-response techniques cannot distinguish between a CFSTR with good micromixing and another with poor micromixing.

2.4. Reaction Kinetics

The rate of reaction is characterized by a *rate equation* which is generally the result of an empirical curve-fitting procedure. The values of the rate constants must be found by experiment, even if the form of the rate equation was suggested by a theoretical analysis or mechanistic model. The determination of a rate equation usually requires a study to determine the influence of concentration, followed by the effects of pH, temperature etc. on the reaction rate coefficients.

2.4.1. The Rate Equation

The *n*th *order rate equation* has been used to great advantage in chemical reactor theory. The equation is a descriptive, two-parameter model and is useful over restricted ranges of experimental data:

$$r = kc^n \tag{14}$$

where r = reaction rate $[ML^{-3}t^{-1}]$
c = reactant or product concentration $[ML^{-3}]$
k = rate coefficient $[t^{-1}m^{1-n}L^{3(n-1)}]$
n = reaction order (dimensionless)

The equation has proven useful because of its limited number of parameters and its simplicity when used in material balance equations.

Another form of rate equation is used extensively in microbial reactors and can be described as the *saturation rate equation*:

$$r = \frac{k_1 c}{k_2 + c} \tag{15}$$

where r = reaction rate $[ML^{-3}t^{-1}]$
c = reactant or product concentration $[ML^{-3}]$
k_1 = rate coefficient $[ML^{-3}t^{-1}]$
k_2 = saturation coefficient $[ML^{-3}]$

From a practical standpoint, the numerical values of the rate coefficients are most important and will depend on the units of measurement. For example, glucose removal rate in a microbial reactor can be described by eqn (15) where c can be expressed as either glucose or organic carbon concentration. When carbon concentration is the measured variable, k_1 and k_2 will be 40% of the values obtained using glucose measurements since glucose is 40% carbon by weight. Similarly, confusion can result when molar concentration units are compared to mass concentration units. Molar units are sometimes awkward in microbial systems when dealing with microbial matter which cannot be readily characterized by a "molecular weight".

2.4.2. Rate Equation Versus Balance Equation

It is critically important to distinguish between a rate equation which describes a transformation and a balance equation which considers both transfer and transformation processes. In a batch reactor, where there are no transfer or transport processes, *the balance equation* for a first-order reaction is expressed as follows:

$$V\frac{dc}{dt} = kcV \qquad (16)$$
$$\underset{\text{rate of accumulation}}{} \quad \underset{\text{process rate}}{}$$

where k = rate coefficient (t^{-1}).

The expression can be confusing, especially if it is referred to as a rate equation. *The rate equation* for this process is correctly expressed as

$$r = kc \qquad (17)$$

The balance equation for the batch reactor can unambiguously be expressed by eqn (16) or as

$$V\frac{dc}{dt} = rV \qquad (18)$$

A material balance for a continuous reactor (Fig. 2) using a similar rate

6. PROCESS ANALYSIS IN MICROBIAL SYSTEMS

expression can be expressed as follows:

$$V\frac{dc}{dt} = F(c_i - c) + rV \tag{19}$$

$$\underbrace{V\frac{dc}{dt}}_{\substack{\text{rate of}\\ \text{accumulation}}} = \underbrace{F(c_i - c)}_{\substack{\text{net rate of}\\ \text{input by flow}}} + \underbrace{kcV}_{\substack{\text{process}\\ \text{rate}}} \tag{20}$$

Equation (20) reduces to the balance equation for the batch reaction (eqn (16)) process if $F = 0$.

2.4.3. Steady State Versus Equilibrium

In batch reaction processes, changes in composition occur with time regardless of whether spatial uniformity exists. The changes occur from moment to moment until thermodynamic equilibrium is reached (or until the reaction is completed). The continuous reaction process is different in that changes in composition occur spatially (e.g. a plug flow reactor or CFSTRs in series). Any part of the system tends toward a time–invariant state.

The movement toward a time–invariant state requires constant feed conditions, constant rate of heat removal etc. Even so, reaction systems have been observed in which concentrations of components oscillate continuously about particular values.

It is important to note that steady state *is not* an equilibrium. The term equilibrium should be reserved for the time–invariant state of closed systems. The steady state of an open system, such as a continuous reactor, depends on the flow rate, reaction rates and the size of the system. In the theory of non-equilibrium or open-system thermodynamics, the time–invariant or steady state in an open system possesses the same meaning as the equilibrium state in equilibrium thermodynamics. In brief, the steady state is the most orderly, efficient and economical state of an open system.

2.4.4. Classification of Reactions

A useful method of classifying reactions is according to the number and types of phases involved. A phase is that part of a system which is physically and chemically uniform throughout. A reaction is *homogeneous* if it takes place in one phase alone. A reaction is *heterogeneous* if it requires the presence of at least two phases to proceed at its characterisic rate. A phase implies uniform composition. Consequently, the choice of classification is sometimes difficult and depends on which description or model is more useful. Rigorously, all microbial reactions are heterogeneous since the biomass constitutes a "solid" phase and the substrate being consumed is soluble, i.e. in the liquid phase. Therefore, segregated models are useful. However, at high substrate concentrations and/or low concentrations of well-dispersed biomass, distributed models are frequently satisfactory and microbial reactions can be considered homogeneous.

In heterogeneous reaction systems, reactants must be transported from one phase to another and the rate of transport may control the overall process rate observed. The transport rate is dependent on mixing intensity and system geometry. In any case, if the process consists of a number of rate processes in series (e.g. transport and then reaction), the slowest step of the series exerts the greatest influence and *controls* the overall rate.

2.4.5. *Expression of Rates*

Process rates are conveniently expressed as intensive measures (i.e. independent of system mass) so that results can be used in analysing systems of varying size. Therefore, the process rate, R, can be expressed per unit reactor volume:

$$r_v = R/V \tag{21}$$

where V = reactor volume (L^3)
R = conversion rate (Mt^{-1})

Certain other forms of expression for rate will be very useful. Conversion rate can be based on unit biomass in a dispersed, well-mixed, system:

$$r_x = R/X \tag{22}$$

where X = biomass in the system (M)
r_x = reaction rate per unit biomass ($MM_x^{-1}t^{-1}$ or possibly t^{-1})

For heterogeneous systems (e.g. biofilms), the area of reaction surface is used for expressing rate.

$$N = R/A \tag{23}$$

where A = area of reactive surface (e.g. biofilm surface area) in the system $[L^2]$
N = flux of reactant at the reactive surface ($ML^{-2}t^{-1}$)

2.4.6. *Reaction Stoichiometry*

Stoichiometry is the application of the law of conservation of mass and the chemical laws of combining weights to chemical processes. In its broadest

6. PROCESS ANALYSIS IN MICROBIAL SYSTEMS

sense, stoichiometry is a *system of accounting* applied to the mass and energy participating in a process involving chemical or physical change. It is a system of calculation which permits a surprisingly large amount of information to be obtained from a seemingly small number of facts.

Stoichiometry provides information concerning the types of changes and maximum extent of changes that can occur in a reaction process. In an abiotic system, thermodynamic calculations allow the determination of the equilibrium constant for a reaction and, hence, the actual yield or reaction products for given conditions. Such calculations require the knowledge of the exact nature of the reactants and the products which are not always determinable in a biotic environment.

In order to define the stoichiometry of a reaction, the stoichiometric coefficients in the reaction equation must be determined. For example, consider the chemical oxidation of glucose:

$$C_6H_{12}O_6 + 6\,O_2 \rightarrow 6\,CO_2 + 6\,H_2O \tag{24}$$

The stoichiometric coefficients for glucose, oxygen, carbon dioxide and water are -1, -6, 6 and 6, respectively. The units of the stoichiometric coefficient are *moles*. More generally, the relationship is expressed by a mass balance equation as follows:

$$\alpha_1 M_1 + \alpha_2 M_2 + \alpha_3 M_3 + \cdots + \alpha_k M_k = 0 \tag{25}$$

where M_i = molecular weight of the ith component ($M\,\text{mol}^{-1}$)
α_i = stoichiometric coefficient (mol)
$\alpha_i \leq 0$ for reactants
$\alpha_i \geq 0$ for products

It follows that the stoichiometric relationship also provides a convenient method for comparing rates of reaction for the various reaction components. If the production rate of component i is $r^{(i)}$, then for the reaction described in eqn (24),

$$r^{(CO_2)} = r^{(H_2O)} = -0.17 r^{(\text{glucose})} = r^{(O_2)} \tag{26}$$

For elementary reactions, the overall process rate, r, is the same regardless of the reaction component measured. Elementary reactions have rate equations which correspond to their stoichiometric equation (e.g. $A + B \rightarrow C$ and $-r_A = kC_A C_B$). Therefore, for a reaction involving k components

$$r = \frac{r^{(1)}}{\alpha_1} = \frac{r^{(2)}}{\alpha_2} = \frac{r^{(3)}}{\alpha_3} = \cdots = \frac{r^{(k)}}{\alpha_k} \tag{27}$$

This relationship indicates that reaction rate can be determined by following the rate of appearance or disappearance of the component that is most easily detectable but which is not necessarily the one of major interest. An example of microbiological interest is the microbial oxidation of glucose. Busch (1971) has examined this process in a batch reactor and has observed the following stoichiometry at the point of glucose depletion:

$$C_6H_{12}O_6 + 2.62 O_2 + 0.64\,NH_3$$
$$\rightarrow 3.21\,CH_{1.8}O_{0.5}N_{0.2} + 2.79\,CO_2 + 4.07\,H_2O \quad (28)$$

or expressed in terms similar to those of eqn (25):

$$3.21\,A_1 + 2.79\,A_2 + 4.07\,A_3 - A_4 - 2.62\,A_5 - 0.64\,A_6 = 0 \quad (29)$$

where A_i = molecular weights of reacting components (M mol^{-1})

$CH_{1.8}O_{0.5}N_{0.2}$ = empirical formula for biomass

The progress of this reaction can be followed by monitoring the concentration of any of the reactants or products. However, oxygen can be measured conveniently, easily and accurately and is frequently used to determine rates of aerobic, microbial reactions. From eqn (28), we see that the ratio of glucose removal rate to oxygen removal rates is approximately 3.06. However, microbial reactions are not elementary reactions and caution is advised when applying *overall process stoichiometry* such as in eqn (28).

When materials react to form products, it is usually easy to decide, after examining the stoichiometry, whether a single reaction or a number of reactions are occurring. When a single stoichiometric equation and single rate equation are chosen to represent the progress of the reaction, it is termed a *single* reaction. When more than one stoichiometric equation is used to describe the observed changes, then more than one kinetic expression is needed to follow the changing composition of the reaction components, which results in a *multiple reaction*. Multiple reactions may be classified simply as follows:

$$A \rightarrow B \rightarrow C \quad \text{series reaction}$$

$$A \begin{matrix} \nearrow B \\ \searrow C \end{matrix} \quad \text{parallel reaction}$$

More complicated schemes are possible.

For a single, elementary chemical reaction, it is sufficient to determine the relative amounts of reactant and product at any one time during the reaction in order to obtain the stoichiometric coefficients. However, for multiple reactions, a more detailed and cautious procedure must be used. Consider the following illustrative example. Nitrification refers to the oxidation of ammonia nitrogen to nitrate nitrogen:

$$NH_4^+ + 2\,O_2 \rightarrow 2\,H^+ + H_2O + NO_3^+ \tag{30}$$

However, the conversion is better described as a multiple reaction as follows:

$$NH_4^+ + 1.5\,O_2 \rightarrow 2\,H^+ + H_2= + NO_2^- \tag{31}$$

$$NO_2^- + 0.5\,O_2 \rightarrow NO_3^- \tag{32}$$

Using the notation from eqn (29), the reaction stoichiometry is as follows:

$$2\,A_2 + A_3 - A_4 - 1.5\,A_5 + A_6 = 0 \tag{33}$$

$$\underline{A_1 \qquad\qquad\qquad -0.5\,A_5 \quad -A_6 = 0} \tag{34}$$

$$A_1 + 2\,A_2 + A_3 - A_4 - 2A_5 \qquad = 0 \tag{35}$$

where A_1 = nitrate ion (M mol^{-1})
A_2 = hydrogen ion (M mol^{-1})
A_3 = water (M mol^{-1})
A_4 = ammonium ion (M mol^{-1})
A_5 = oxygen (M mol^{-1})
A_6 = nitrite ion (M mol^{-1})

The course of nitrification in a batch or plug flow reactor is described in Fig. 5. At any time during the reaction, the amount of NH_4^+ reacted results in production of NO_2^- and NO_3^- in a proportion that depends not only on the stoichiometry, but also on the rate of the reactions.

2.4.7. Microbial Reactions Versus Chemical Reactions

The methods for determining the stoichiometric and kinetic parameters for a given conversion have been developed primarily for abiotic reactions. Biochemical conversions mediated by viable organisms differ from abiotic chemical conversions in several important ways:

(1) Microbial reactions are *irreversible*. The stoichiometric end-point generally refers to the exhaustion of one of the reactants.
(2) All microbial reactions are *heterogeneous*.

(3) Relatively low concentrations of reactants and products, in addition to the heterogeneous characteristic, increase the potential for *mass transfer limitations*.
(4) The reaction, frequently represented by one stoichiometric equation, consists of many enzymatic reactions (in series and parallel) occurring in the metabolic region of the cell.
(5) The reactions are generally *autocatalytic*, i.e. biomass and related enzymes increase as the reaction proceeds.
(6) One of the reactants, biomass, has a *structure* and a *history* which can influence the stoichiometry and kinetics of the reaction.

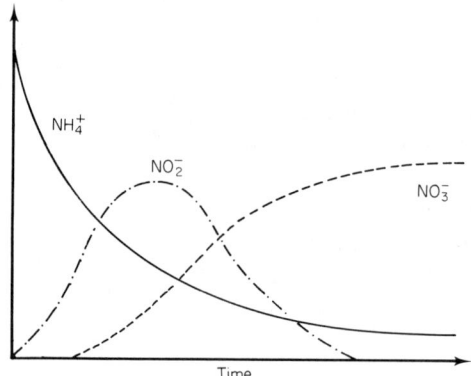

Fig. 5. Progress of the nitrification reactions in a batch reactor.

3. Modelling of Microbial Processes

The process of mathematical modelling is an iterative one in which careful study of a simple model leads to a more realistic, but more complex, model, which is itself the precursor of a better model. The test of each model requires experiments in which the model variables are measured quantities. Hence, it is important to determine which variables should be considered and which variables can be measured in the modelling of microbial processes. The variables of concern can be divided into two categories: *biotic* and *abiotic* variables.

6. PROCESS ANALYSIS IN MICROBIAL SYSTEMS

3.1. Modelling Variables (after Frederickson *et al.*, 1970)

The biotic variables of importance depend on the complexity of the microbial process. If the process involves a monoculture, number of individuals (population density), biomass (or biomass concentration) and distribution of physiological states within the population become quantities of significant interest. If the process involves mixed cultures, then an additional class of variables which describe community structure must be considered. Such quantities include numbers and biomass of the various populations of the community, in addition to the distribution of physiological states within each population. The physiological state is quite important in modelling any microbial process and is somewhat difficult to quantify.

Abiotic variables describe the physical and chemical environment in which the microbial process is occurring. Abiotic variables describe the *state* of the abiotic phase or phases of the reaction system. The *chemical state* describes the concentrations of various chemical substances in the liquid phase such as substrate, nutrients, dissolved oxygen and pH. Another class of variables describes the *physical state* of the system which frequently exerts control on the rate of the biological and chemical processes occurring in the reactor. For example, dilution rate is an important variable controlling microbial processes in a continuous fermentor. Another example is mechanical shear which may alter the physiological state of the organisms either directly by influencing lysis rate or indirectly by influencing mass transfer rates. Exposed cellular surface area per unit reactor volume may significantly influence microbial process rates.

3.2. Physical Principles

As described earlier, mathematical models are generated in part by application of well-established principles of physical and biological sciences. Useful physical principles may be divided into three categories: conservative, thermodynamic and constitutive. These principles have been discussed previously.

3.2.1. Conservation and Thermodynamics

Conservation principles are essentially accounting principles. For any system, the conservation principle for a specific component may be expressed verbally as follows:

$$\begin{matrix} \text{component rate} \\ \text{of accumulation} \\ \text{in the system} \end{matrix} = \begin{matrix} \text{net rate of component} \\ \text{transfer into the system} \\ \text{across its boundaries} \end{matrix} + \begin{matrix} \text{net rate of} \\ \text{component generation} \\ \text{within the system} \end{matrix} \quad (36)$$

Components of importance to microbial process systems subject to this principle include mass, chemical and atomic species, and energy. Conservation includes the first law of thermodynamics which is the principle of conservation of energy. The second law of thermodynamics places restriction on process efficiency and the extent to which these processes can occur. Micro-organisms are so-called open systems, i.e. they are not in thermodynamic equilibrium since they exchange matter and energy with their environment. Hence, a macroscopic description of their behaviour demands the formulation of thermodynamics of irreversible processes.

3.2.2. *Constitution*

Constitutive principles deal with the *rates* of physicochemical processes. Conservation and thermodynamic principles do not deal with mechanism, but constitutive principles do; constitutive principles deal with mechanisms as influenced by the constitution of matter. Examples of constitutive principles are the laws that govern rates of transport (transport phenomena) and rates of chemical reactions (reaction kinetics).

Since conservation principles do not depend on the constitution of matter, a mathematical model expressing a conservation principle is "correct" if all flows, sources and sinks have been included. The usefulness of the model, however, depends primarily on two things:

(1) The choice of the system.
(2) Availability of constitutive principles for description of unknown quantities which appear in the model (e.g. reaction rates).

Since biological mechanisms and biological constitutions are often poorly understood, constitutive relations present the most difficulty in developing mathematical models of microbial processes.

3.3. Biological Principles

Biological principles are not as clearly defined. However, there are at least four that are of general importance in model building.

3.3.1. *Phenotype and Genotype*

The first principle is physiological and ecological and states that the activities of an organism (phenotype) and rates at which these activities are conducted are dependent not only on the organism (genotype) but also on the organism's environment. A corollary of this principle is that the constitution or state of the environment depends upon the activities of organisms contained in it. This principle and its corollary are explicitly incorporated in most microbial process models. For example, the concept of *limiting nutrient* describes the influence of the environment on the organisms (the limiting nutrient controls metabolism) and the influence of the organism on the environment (the organism consumes the limiting nutrient thus altering the state of the environment).

3.3.2. Past History

The second principle is drawn from ecology and genetics. The current phenotype (constitution or state) of an organism depends not only on its genotype but also on the past history of environments experienced by the organism. Thus, cultivating an organism in a changing environment (e.g. batch culture), one may observe differing function and structure within the same genotype. Although this principle is recognized, it is seldom incorporated satisfactorily in microbial process models. If this principle is incorporated, the model is said to be *structured*; otherwise it is unstructured.

3.3.3. Taxonomic Classification

The third principle states that organisms can be classified based on their morphology, growth form and mode of reproduction. Thus, Eubacteriales are unicellular, exhibit intercalary growth and generally reproduce by binary fission and may be distinguished from Actinomycetales which may be filamentous, multicellular, exhibit apical growth and reproduce in several different ways. The implications of this principle are frequently not realized when modelling microbial processes. For example, the same model is frequently used for describing growth of Eubacteriales and Actinomycetales.

3.3.4. Mutation

The fourth principle is the mutability of organisms. Although universally recognized, the principle is rarely considered in modelling population dynamics.

There may be other biological principles of more or less interest but the four listed are generally recognized, but frequently ignored, by the model builders.

3.4. Simplifying Assumptions

The growth of even a single population of micro-organisms is a tremendously complicated process. The task of mathematically modelling such growth requires simplifying assumptions. These introduce inexactitudes into the models since the models will not describe all of the various facets of growth that occur. However, the models will probably describe some facets of growth considered important; only continued experimental testing will lead to progress in mathematical modelling.

Many models of microbial growth make certain simplifying assumptions without stating such assumptions explicitly. It seems important that the more common hypotheses be stated more explicitly. Hence, four will be considered in this section.

3.4.1. *Neglect of the Distribution of States*

Biological populations are composed of individual organisms, and the individuals are physiologically, morphologically and genetically different. Individuals differ in size, shape, staining properties and perhaps in other characteristics, such as motility. Examination for a period of time teaches that individuals differ in "age", if we define a cell's age as the chronological time since it was formed by fission of its parent, since fissions are never fully synchronous events. More refined analyses reveal even more fundamental, if less obvious, distinctions between individuals. In summary, individuals of a population do not all exist in the same "state", but rather represent a distribution of states.

The first simplifying assumption in most microbial growth models is that the distribution of states can be ignored and the properties of the culture can be adequately described in terms of a "typical" individual whose behaviour represents an average over the distribution of states. Such assumptions at once lead to uncertainty about the validity of the model, since they imply that a whole host of parameters of the population are not important in determining the properties or activities of the population.

There are three reasons for making these assumptions; one involves a conceptual difficulty and the other two involve practical difficulties. The conceptual difficulty is: What do we mean by the "state" of an individual organism? Can we use some obvious parameter associated with a cell as a measure or index of its state? Would the age of a cell (as defined above) serve this purpose? Or can we use the size of a bacterial cell as an index of its state? Or is some more general notion of state, such as the biochemical constitution of the cell needed?

An obvious practical difficulty involved in the use of any model that recognizes a distribution of states is: How does one *measure* the states of individual organisms so that some idea of the distribution can be obtained? The idea of bacterial size as an index of state is attractive here, since electronic devices measure size distributions quickly and accurately. Many workers might prefer to use age or some other cell feature as an index of

6. PROCESS ANALYSIS IN MICROBIAL SYSTEMS   199

state. The theoretical work of Frederickson *et al.* (1967) has shown that any *single* index of state is going to be inadequate in all but the most rigidly controlled growth conditions.

The other practical difficulty involved here is the fact that the equations resulting from models recognizing a distribution of states quickly became *mathematically intractable*.

For these reasons, only models where distribution of states among individual organisms is neglected will be considered.

3.4.2. *Neglect of Segregation*

In unicellular micro-organisms, life is *segregated* into structurally and functionally discrete units—cells. Hence, the *number* of individual organisms present in a population must be an important parameter for the description of the population. Such quantities as the biomass of the population must also be involved. Nevertheless, number must be a quantity of prime importance since the biological characteristics of a population composed of $2n$ organisms having total biomass M are not the same as those of a population composed of n organisms having total biomass M. ($2n$ and n as used here refer to number of organisms.)

In spite of the foregoing arguments, many models make the assumption that segregation of life into discrete units can be ignored. With such an assumption, number of organisms is not admitted as a parameter to be described by the model and, in effect, the model views the population as biomass *distributed continuously* throughout the culture. Models based on such neglect of segregation are *non-segregated* or *distributed*.

Can a non-segregated model have any success in the description of unicellular growth? There may be no *practical* need for knowing numbers of organisms present (biomass may be the quantity of practical importance), but in general, increase in number (proliferation) and increase in biomass (growth) are coupled processes so that one cannot really omit the one from a model purporting to describe the course of the other. A possible explanation for the success of some non-segregated models is that they have been applied to *balanced growth*, or nearly balanced growth situations. Under these conditions, growth and proliferation are *proportional*, so that biomass is directly related to number of organisms.

3.4.3. *Neglect of Stochastic Phenomena*

It is not possible to predict the behaviour of individual microbial cells with certainty. Thus, generation times of individual bacterial cells are not all the same, but rather show random deviations about mean values. Models for microbial propagation generally make the simplifying assumption that the foregoing stochastic phenomena can be neglected, and that growth can be treated as a deterministic process. If this assumption cannot be made, then non-segregated models cannot be used, either; if cells divide at random times, the number of cells present must be a variable of the model.

Stochastic population models (so-called "birth-and-death processes") very quickly lead to formidable mathematical difficulties, even when one attempts to model only very simple biological phenomena. Hence, it is desirable to avoid such models whenever possible.

Fortunately, avoidance of these models is usually permissible in microbial processes because one is dealing with an enormous number of cells so that random deviations cancel out. There are situations where random deviations are important, however, and these always deal with cases where the *total number* of cells involved is small. Sterilization is one such situation and various transient growth situations, in which the population size for one reason or another becomes quite small (such as near the critical dilution rate in continuous culture) and others. Hence, stochastic phenomena will be ignored.

3.4.4. *Neglect of Biological Structure*

Two micro-organisms having the same biomass and inhabiting the same environment may nevertheless have widely different properties and activities. This is the problem of *state* again; the two organisms have different states. If the model recognizes the existence of a distribution of states, it should also recognize that distribution may change in response to changes in the environment. Or, if the model does not recognize the existence of a distribution of states, it should at least recognize the possibility that the state of the average or typical organism (which is all such models consider) can change in response to changes in the environment. This means that parameters *in addition* to population number and population biomass must in general be important for the description of population behaviour.

Many microbial process models currently used do not recognize this; in most of these models, population biomass is the sole variable employed for describing the population. Since this procedure regards organisms and population biomass as featureless, structureless entities, we shall call such models *unstructured*.

4. Biofilm Processes

Development of a systematic understanding of biofilm processes has been limited because of the interaction of several contributing rate processes.

Mechanistically, biofilm development may be described as the net result of the following:

(1) *Transport of material from the bulk fluid to the surface and attachment to the surface.* Materials can be soluble (microbial nutrients and organics) or particulate (viable micro-organisms, their detritus or inorganic particles). Also, suspended particles of sufficient mass may control films by "scouring" action.
(2) *Microbial metabolism within the film.* Microbial growth in the biofilm and extracellular polymers produced by the micro-organisms contribute to the biofilm deposit and promote adherence of inorganic suspended solids.
(3) *Fluid shear stress at the surface of the film.* Such forces can limit the overall extent of the biofilm deposit by re-entraining attached material.
(4) *Surface material and roughness.* Surface properties can influence micromixing near the surface and corrosion processes. Some metal surfaces may release toxic components into the biofilm inhibiting growth and/or attachment. Some surfaces are porous (e.g. wood) and provide environments protected from fluid shear forces.
(5) *Biofilm control procedures.* In heat exchangers and other systems where biofilms interfere with performance, chemical compounds are introduced to control or prevent them. Chlorine, the most commonly used chemical, oxidizes biofilm polymers causing disruption and partial removal. Inactivation of a portion of the microbial population also occurs. Altered biofilm "roughness" and decreased viable cell numbers will influence "regrowth" rates of the biofilm. Mechanical forces can physically remove portions of the attached film.

The physical, chemical and biological structure of the biofilm is influenced by its environment and, in turn, influences its environment (e.g. substrate removal, fluid frictional resistance and heat transfer resistance).

4.1. Properties and Composition of Biofilms

Micro-organisms, primarily bacteria, adhere to surfaces ranging from the human tooth and intestine to the metal surface of condenser tubes exposed to turbulent flow of water. The micro-organisms "stick" by means of extracellular polymer fibres, fabricated and oriented by the cell, that extend from the cell surface to form a tangled matrix termed a "glycocalyx" by Costerton *et al.* (1978). The fibres may conserve and concentrate extracellular enzymes necessary for preparing substrate molecules for ingestion, especially high molecular weight or particulate substrate frequently found in natural waters.

The biofilm surface is highly adsorptive, partially due to its polyelectrolyte nature, and can collect significant quantities of silt, clay and other detritus in natural waters.

Physical, chemical and biological properties of biofilms are dependent on the environment to which the attachment surface is exposed. The physical and chemical microenvironment combine to select the prevalent micro-organisms which, in turn, modify the microenvironment of the surface. As colonization proceeds and a biofilm develops, gradients develop within the biofilm and average biofilm properties change. Changes in biofilm properties that occur during biofilm development must be considered when attempting to predict the effect of biofilms on mass, fluid and heat transport in turbulent flow systems. These changes have been largely ignored in past studies.

4.1.1. *Physical Properties*

Relevant *thermodynamic properties* of biofilm are its volume (thickness) and mass. In turbulent flow systems, wet biofilm thickness (Th) seldom

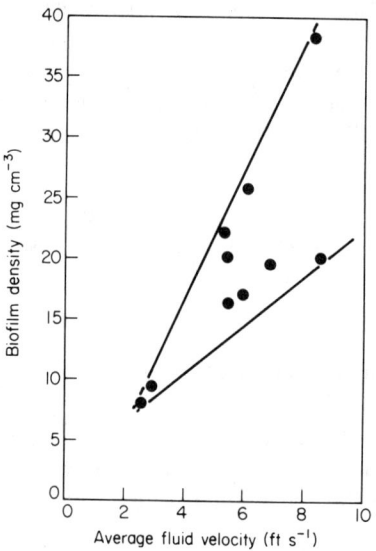

Fig. 6. Influence of fluid shear stress of biofilm density.

6. PROCESS ANALYSIS IN MICROBIAL SYSTEMS

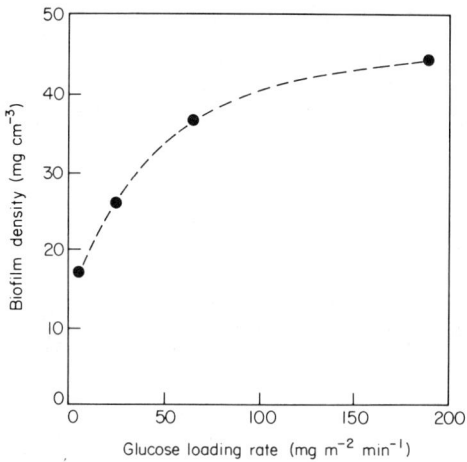

Fig. 7. Influence of glucose loading rate on biofilm density.

exceeds 1000 μm (Zelver, 1979). The biofilm dry mass density (ρ_{Th}) can be determined from the wet biofilm thickness if the biofilm mass per unit area ($\rho_{Th} Th$) is known. ρ_{Th} reflects the attached dry mass per unit wet biofilm volume and measured values in turbulent flow systems range from 10 to 50 mg cm^{-3}. ρ_{Th} increases with increasing turbulence (Characklis, 1980) and increasing substrate loading (Trulear and Characklis, 1982) as indicated in Figs 6 and 7. The increase in ρ_{Th} with increasing turbulence may be caused by one of the following phenomena:

(1) Selective attachment of only certain microbial species from the available population.
(2) Microbial metabolic response to environmental stress.
(3) Fluid pressure forces "squeeze" loosely bound water from the biofilm.

The relatively low biofilm mass densities compare well with observed water content of biofilm (Characklis et al., 1981; Characklis, 1973, 1980).

The *transport properties* of biofilm are of critical importance in quantifying effects of biofilms on mass, heat and momentum transfer. Diffusion coefficients for various compounds through microbial aggregates have been reported in the literature, mostly for floc particles (Table 4). Matson and Characklis (1976) report variation in the diffusion coefficient for glucose and oxygen with growth rate and carbon-to-nitrogen ratio. In biofilms, the

Table 4. Experimental diffusion coefficient measurements from the literature (from Matson and Characklis, 1976).

Reactant	Diffusivity (10^{-5} cm^2 s^{-1})	D_{floc}/D_{H_2O} ×100%	Biomass type	Growth system	Procedure	Reference
Oxygen	1.5	70	Bacterial slime	Rotating tube	Reaction products analysis	Tomlinson and Snaddon, 1966
Oxygen	0.21	8	Fungal slime	Fluidized reactor	Non-linear curve fit	Mueller et al., 1966
Glucose	0.048	8	*Zooglea ramigera*	Fluidized reactor	Non-linear curve fit	Baillod and Boyle, 1970
Glucose	0.06–0.6	10–100	*Zooglea ramigera* Mixed culture	Fluidized reactor	Two chamber	Pipes, 1974
Oxygen	2.2	90	Nitrifier culture	Fluidized reactor	Two chamber	Williamson and McCarty, 1976
Ammonia	1.3	80				
Nitrate	1.4	90				
Oxygen[a]	0.4–2.0	20–100	Mixed culture	Fluidized reactor	Two chamber	Matson, 1975
Glucose[a]	0.06–0.21	10–30				Matson, 1975

[a] Tests conducted under a variety of experimental conditions.

Table 5. Viscoelastic properties of biofilm developed at 40°C at a fluid shear stress of 3.3 N m^{-2}. Glucose was growth-limiting and was applied at 6.2 mg m^{-2} min^{-1} (Characklis, 1980).

Elastic (storage) modulus	59.5 N m^{-2}
Viscous (loss) modulus	118 N m^{-2}

diffusion coefficient is most probably related to biofilm density. *In situ* rheological measurements indicate that the biofilm is viscoelastic with a relatively high viscous modulus as indicated in Table 5. Reported biofilm thermal conductivities are presented in Table 6. As expected from reported water content, biofilm thermal conductivity is not significantly different from water.

Table 6. Thermal conductivity of biofilm and other selected materials relevant to biofouling of heat exchangers.

Material	Thermal conductivity (W m^{-1} K^{-1})	Temperature (°C)	Reference
Biofilm	0.68 ± 0.27	28.3 ± 0.3	Characklis, et al., 1981
	0.71 ± 0.39	26.7 ± 0.3	
	0.57 ± 0.10	28.3 ± 0.3	
Water	0.61	26.7	Weast, 1973
	0.62	28.3	
Carbon steel	51.92	0–100	Perry and Chilton, 1973
Steel	46.86	18	Atomic Energy Commision, 1955
Stainless steel (type 316)	16.30	0–100	Perry and Chilton, 1973
Aluminium 5052	138.46	20	Perry and Chilton, 1973
	205.85	100	
Cupronickel 10% 706	44.71	0–100	Perry and Chilton, 1973
Copper	384	18	Atomic Energy Commission, 1955
Titanium (commercial pure)	16.44	0–100	Perry and Chilton, 1973
Glass	0.6–0.9	—	Weast, 1973

Table 7. Chemical properties of biofilms obtained from fouled surfaces experiencing excessive frictional losses.

	Pollard and House, 1959	Minkus, 1954	Minkus, 1954	Arnold, 1936	Characklis, 1980
Water	87	85.6	90	95	96
Volatile fraction	2.5	2.7	1.9	2.4	3.2
Fixed fraction	10.5	11.7	8.1	2.6	0.8
Si (as percentage fixed fraction)		7.0	11.8	12.5	
Fe (as percentage fixed fraction)		18.5	7.9	1.4	
Al (as percentage fixed fraction)		7.5		3.9	
Ca (as percentage fixed fraction)		1.0			
Mg (as percentage fixed fraction)		2.5	5.6	3.2	
Mn (as percentage fixed fraction)		59.5	56.3	4.9	

Table 8. Chemical composition of biofilms obtained in the field and laboratory emphasizing the primary constituents (C, N, P).

	Percentage dry weight					
Source	Carbon (C)	Nitrogen (N)	Phosphorous (P)	Fixed solids	C/N	C/P	Reference
Biofilm— power plant condenser	6.4–13.8	0.5–3.0	—	—	2–27	—	Anderson et al., 1977
Biofilm— laboratory reactor	42.8	10.0	—	—	4.3	—	Kornegar and Andrews, 1967
Biofilm— laboratory reactor	19.0	9.2	1.8	20	2.1	10.5	Characklis, 1980
E. coli	50.0	14.0	3.0	—	3.6	16.7	Gunsalus and Stanier, 1960

4.1.2. Chemical Properties

Inorganic composition of biofilms undoubtedly varies with the chemical composition of the bulk water and probably affects the physical and biological structure of the film. Calcium, magnesium and iron affect intermolecular bonding of biofilm polymers which are primarily responsible for the structural integrity of the deposit. In fact, EDTA is effective in detaching biofilm (Characklis, 1980). In heat exchangers, corrosion products and inert suspended solids can adsorb to the biofilm matrix and influence its chemical composition. Table 7 reports the range of inorganic composition observed in selected biofilms.

The organic composition of the biofilm is strongly related to the energy and carbon sources available for metabolism. Classic papers (Herbert, 1961; Schaechter et al., 1958) have demonstrated the effect of environment and microbial growth rate on the composition of the cells and their extracellular products. For example, nitrogen limitation can result in production of copious quantities of microbial extracellular polysaccharides. Trulear (1983) indicates that only 10% of the biofilm organic carbon can be attributed to the cells in a balanced growth media where glucose is the limiting growth factor. Table 8 presents data on the composition of biofilms developed in the field and in the laboratory. In terms of macromolecular composition, Bryers (1979) has measured protein-to-polysaccharide mass ratios ranging from 0 to 10 (polysaccharide concentration in terms of glucose and protein concentration based on casein) with increasing biofilm accumulation. Other chemical analyses of biofilm have been reported by Bryers and Characklis (1979).

4.1.3. Biological Properties

The organisms which colonize the attachment surface will strongly influence biofilm development rate and biofilm chemical and physical properties. However, organism–organism and organism–environment interactions undoubtedly shift population distributions during biofilm accumulation. Several investigators have observed succession during biofouling (Marshall, 1976; Corpe, 1978).

The first visible signs of microbial activity on a surface are usually small "colonies" of cells distributed randomly on the surface. As biofilm development continues, the colonies grow together forming a relatively uniform biofilm. The viable cell numbers are relatively low in relation to the biofilm volume (10^4–10^8 cm^{-3} biofilm) occupying only from 1 to 10% of the biofilm in dilute nutrient solutions (Characklis, 1980). Jones et al. (1969), present photomicrographs which corroborate these data in natural and laboratory systems.

In many cases, filamentous forms emerge as the biofilm develops further. *Hyphomicrobium*, *Sphaerotilus* and *Beggiatoa* are frequently identified. The filamentous forms may gain an ecological advantage as the biofilm develops since their cells can extend into the flow to obtain needed nutrients or oxygen which may be depleted in the deeper portions. Jenkins (1980) has demonstrated the competitive advantage of a *Sphaerotilus* sp. over a *Pseudomonas* sp. under low oxygen tension in a fermenter. Trulear (1983) has demonstrated the advantage obtained by *Sphaerotilus* over *Pseudomonas* in a reactor with a high surface area-to-volume (approx. 3 cm^{-1}).

4.2. Rate Processes Contributing to Biofilm Development

In this discussion, biofilm development will be considered to be the net result of the following physical, chemical and biological processes:

(1) Transport of organic molecules and microbial cells to the wetted surface (Fig. 8).
(2) Adsorption of organic molecules to the wetted surface resulting in a "conditioned" surface (Fig. 9).
(3) Adhesion of microbial cells to the "conditioned" surface (Fig. 10).
(4) Metabolism by the attached microbial cells resulting in more attached cells and associated material (Fig. 11).
(5) Detachment of portions of the biofilm (Fig. 16).

4.2.1. Transport to the Wetted Surface

When a clean surface is immersed in natural water, transport controls the initial rate of deposition (Fig. 8). In very dilute suspensions of microbial cells and nutrients, transport of microbial cells to the surface may be the rate-controlling step for long periods of time. Biofilm development in open ocean waters or distilled water storage tanks may be illustrative of these cases. Transport of molecules and particles smaller than 0.01–0.1 μm is described satisfactorily in terms of diffusion. In turbulent flow, the diffusion equation must be modified to include turbulent eddy transport. Transport of such small molecules and particles is relatively rapid compared to transport of larger particles. Consequently, adsorption of organic molecules is reported to occur "instantaneously" in many cases as schematically illustrated in Fig. 9.

Larger particles develop a sluggishness with respect to the surrounding fluid. As the particle approaches the wetted surface, eddy transport diminishes and the viscous sublayer exerts a greater influence. For soluble matter and very small particles, diffusion can adequately describe transport

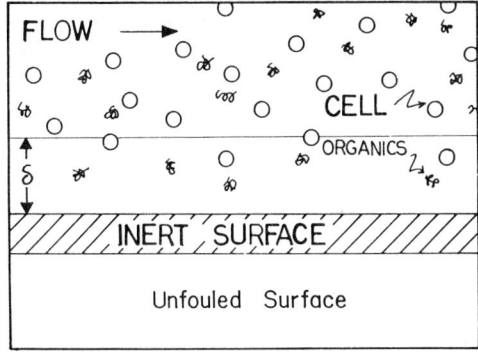

Fig. 8. A "clean" surface exposed to a turbulent flow of fluid containing dispersed micro-organisms, nutrients and organic macromolecules. δ refers to the viscous sublayer thickness.

in the viscous sublayer (Lister, 1981; Lin *et al.*, 1953; Wells and Chamberlain, 1967). For larger particles, other mechanisms must be considered to explain experimental observations.

Within a turbulent flow regime, larger particles suspended within the fluid are transported to the solid surface primarily by fluid dynamic forces. Particle flux to the surface increases with increasing particle concentration. However, particle flux is also strongly dependent on the physical properties of the particles (e.g. size, shape, density) and is influenced by many other forces near the attachment surface.

Microbial cells (0.5–10.0 μm effective diameter) can be transported from

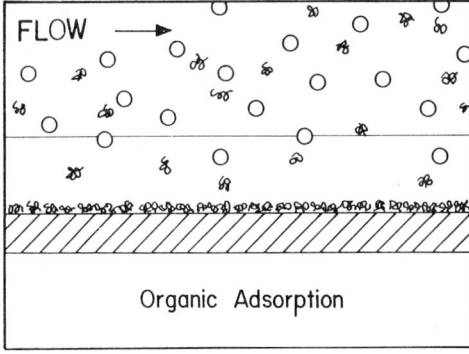

Fig. 9. Transport and adsorption of an organic monolayer on a clean surface.

the bulk fluid to the wetted surface by several processes including the following:

diffusion (Brownian),
gravity,
thermophoresis,
taxis,
fluid dynamic forces,
  inertia,
  lift,
  drag,
  drainage,
  downsweeps.

Particles in turbulent flow are transported to within short distances of the surface by eddy diffusion. Particles are propelled into the viscous (or laminar) sublayer under their own momentum. Turbulent eddies supply the initial impetus and frictional drag slows down the particle as it penetrates the viscous sublayer (Friedlander and Johnstone, 1957; Beal, 1970). For microbial cells, the inertial forces are very small because of their small diameter and density (in relation to water).

If the particle is travelling faster than the fluid in the region of the wall, the *lift force* directs the particle towards the wall (Rouhiainen and Stachiewicz, 1970). This would normally be the case if particle density is greater than fluid density and the particle is moving towards the wall. *Frictional drag forces* can be significant, especially in the viscous sublayer region. The drag force slows down the particle as it approaches the surface and is proportional to difference between particle velocity and fluid velocity.

If the mass density of the particle, ρ_p, differs substantially from the fluid density, ρ, the *gravity force* may be significant. For microbial cells in turbulent flow, the gravity force is generally negligible. *Thermophoresis* is only relevant when particles are being transported in a temperature gradient (Lister, 1981). If the surface is hot and the bulk fluid is cold, the thermophoretic force will repel the particle from the surface. *Eddy diffusion* may be instrumental in dispersing particles in the turbulent core region, thus maintaining a relatively uniform concentration in that region. However, eddy diffusion will not be significant in transporting particles to the wall. *Brownian diffusion* contributes little to the transport of microbial cells (>1.0 µm diameter) in turbulent flow. Certain microbes are capable of *motility* or *taxis* through their own internal energy. Velocities as high as 4×10^{-4} cm s^{-1} have been observed. Taxis could possibly be a significant transport process within the viscous sublayer. For particles in liquids, the *fluid drainage force* is significant (Lister, 1981). The drainage force describes the resistance the particle encounters near the wall due to the pressure in the draining fluid film between the two approaching surfaces. This force is quite large for a microbial cell as it approaches the wall.

Recent published research on the structure of the viscous sublayer in turbulent flow indicates that *"downsweeps"* of fluid from the turbulent core penetrate all the way to the wall (Cleaver and Yates, 1975, 1976). Particles in the bulk fluid are transported all the way to the wall by these convective downsweeps. Aside from lift, this is the only fluid mechanic force directing the particle to the wall. Downsweeps are apparently quite important in terms of particle transport to the wall in turbulent flow.

For a Reynolds number = 30 000 in tube flow, the bursts resulting from the downsweeps have the following characteristics:

Burst diameter	0.11 cm
Average axial distance between bursts	0.50 cm
Mean time between bursts	0.006 s

Minimum transport rate of particles would be observed when particle diameter approximates 0.1×10^{-4} cm under constant fluid flow conditions. At this diameter, *Brownian diffusion* starts exerting a significant effect. Calculated particle flux in the pipe for Reynolds number = 30 000 and a bulk fluid particle concentration 10^4 particles cm^{-3} is approximately 0.1 particles $cm^{-2} s^{-1}$.

Surface roughness significantly influences transport rate and microbial cell attachment for several reasons including the following:

(1) Increases convective mass transport near the surface.
(2) Provides more "shelter" from shear forces for small particles.
(3) Increases surface area for attachment.

If surface roughness elements are larger than the viscous sublayer, the roughness can be measured quantitatively by hydraulic methods. If surface roughness elements are smaller than the viscous sublayer (i.e. microroughness), measurements of roughness are difficult to quantify and interpret. Brown (1974) reports that particle deposition from gases is very sensitive to roughness too small to be detected by fluid frictional resistance.

When a "clean" surface first contacts water with biological activity, organic substances and microbial cells must be transported to the surface before biofilm development can begin. Consequently, the rate of transport determines the length of the "induction" period, i.e. the initial period during which no macroscopic effects of the biofilm are evident. In very dilute solutions (e.g. open ocean), the rate of transport may control the overall rate of biofilm development for long periods. Rate of transport is proportional to the concentration difference between the bulk fluid and the surface. In dilute solutions, this difference is small. The flow regime (zero, laminar or turbulent) also significantly influences transport rates and should be defined carefully in any experimental system used for biofilm studies. Surface characteristics are also critical to the repeatability and applicability of the results because a rough surface will increase transport and attachment rates. Which rate controls—rate of transport or rate of adhesion?

4.2.2. Adsorption of Organic Molecules to the Wetted Surfaces

Figure 8 illustrates an initially "clean" surface exposed to turbulent flow of a fluid containing dispersed micro-organisms, nutrients and organic macromolecules. Micro-organisms select their habitats on the basis of many factors, including the nature of the wetted surface (material of construction and surface roughness). Adsorption of an organic monolayer occurs within minutes of exposure as shown in Fig. 9 and changes the properties of the wetted surface. Investigations have shown that materials with diverse surface properties (e.g. wettability, surface tension, electrophoretic mobility) are rapidly conditioned by adsorbing organics when exposed to natural waters with low organic concentrations. These organic molecules are usually polysaccharides or glycoproteins. Loeb and Neihof (1975) and DePalma et al. (1979) have measured adsorption rates of organic molecules in seawater, and Bryers (1979) has observed adsorption rates in a laboratory system. Rates and extent of adsorption in these investigations are presented in Table 9. Maximum accumulation from molecular fouling is less than 0.1 μm. The rate of molecular fouling can be considered instantaneous since it is much greater than the rate of microbial fouling. Based on "thickness" measurements, molecular fouling can have no significant effect on fluid flow or heat transfer. Nevertheless, the surface properties resulting from adsorption of an organic film may affect the sequence of microbial events which follow.

Costerton et al. (1978) have discussed the pronounced specificity of some bacteria that attack only a particular animal host tissue and suggest that specificity may be explained by the specificity of the host-tissue glycocalyx. It remains to be seen whether an abiotic surface, wetted by the adsorption of organic molecules indigenous to that environment, will be initially colonized by a *specific* microbial cell.

Brash and Samak (1979) present experimental evidence that significant turnover occurs in molecular (proteinaceous) films developed on polyethylene. Protein molecules in the bulk fluid are continuously exchanging with adsorbed proteins. This suggests that dispersed microbial cells in the

Table 9. Maximum rate and extent of molecular fouling.

Maximum rate (nm min^{-1})	Maximum accumulation (nm)	Maximum accumulation (μg COD cm^{-2})	Surface	Reference
0.15–0.45	30–80		Pt^a	Loeb and Neihof, 1975
0.004	7.1		Ge^b	DePalma et al., 1979
0.004	77.3		Ti^b	
0.01^5	13.5^c	1.5	glass^d	Bryers, 1979
0.22^5	22.5^c	2.5	glass^e	

^{a} Immersed in quiescent Chesapeake Bay water (3–4°C) containing 2.3 mg carbon l^{-1}, salinity between 9 and 16‰ and pH between 7.9 and 8.2.
^{b} Gulf of Mexico water (22°C) flowing past the surface at a fluid shear stress of 7.1 N m^{-2}. Salinity was 34‰. Carbon concentration not reported.
^{c} Estimated from measurements of chemical oxygen demand (COD) adsorbed per unit area. Assumed DOC of protein is 0.855 mg COD mg^{-1} protein and protein density is 1.3 g protein cm^{-3}.
^{d} Medium consisted of sterile 1:1 w/w trypticase soy broth–glucose mixture (34°C; pH 8). The glass surfaces were immersed in tubes placed in a mechanical shaker. Carbon concentration was approximately 80 mg carbon l^{-1}.
^{e} Medium was effluent (30°C; pH 8) from a chemostat (10–20 mg l^{-1} COD, 3 mg l^{-1} polysaccharide) with no primary substrate remaining. Micro-organisms were present (approximately 10^6 cells ml^{-1}) but no cells attached during the period of interest. Fluid shear stress was 3.8 N m^{-2}.

bulk fluid and their associated extracellular material may be continually exchanging with biofilm material at the wall.

4.2.3. Adhesion of microbial cells to the wetted surface

Previous research (Marshall et al., 1971; Zobell, 1943) suggests the existence of a two-stage adhesion process: (1) reversible adhesion followed by, (2) an irreversible adhesion. Reversible adhesion refers to an initially weak adhesion of a cell which no longer exhibits Brownian motion but is readily removed by mild rinsing. The adhesive forces which hold the cell at the wall during reversible adhesion probably include electrostatic, London–van der Waals, interfacial tension and covalent bonding. Conversely, irreversible adhesion is a permanent bonding to the surface, usually aided by the production of extracellular polymers. Cells attached in this way can only be removed by rather severe mechanical treatment. Marshall (1976) and Corpe (1970) have implicated polysaccharides and glycoproteins in irreversible adhesion (Fig. 10).

Most of the research on cell adhesion has been conducted at very low fluid shear stress or in quiescent conditions (Fletcher, 1977). Under these

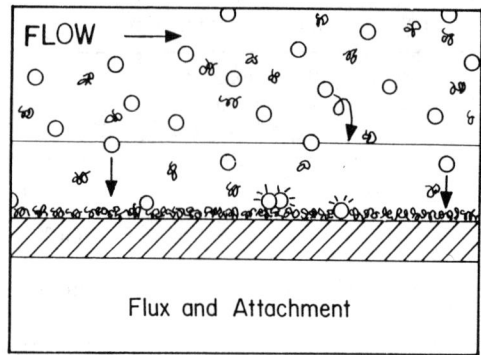

Fig. 10. Transport and attachment of microbial cells to the conditioned surface.

conditions sedimentation or diffusion may control the rate of adhesion. There is yet to be a demonstration of reversible adhesion in turbulent flow.

In turbulent flow, the *net* rate of adhesion is the quantity most easily measured. The net rate of adhesion is the difference between the rate of adhesion and rate of detachment. Detachment results from several forces including fluid dynamic forces, shear forces, lift (upsweeps) and taxis. Upsweeps are analogous to the downsweeps discussed previously. Upsweeps result in turbulent bursts which move away from the surface into the bulk flow. Upsweeps generate a lift force normal to the surface which can influence detachment. Drag or viscous shear forces act in the direction of flow on attached cells and are approximately 1000 times greater than the lift forces acting on attached cells. Note that although viscous shear may dislodge a particle, unless a lift force is present, the particle will

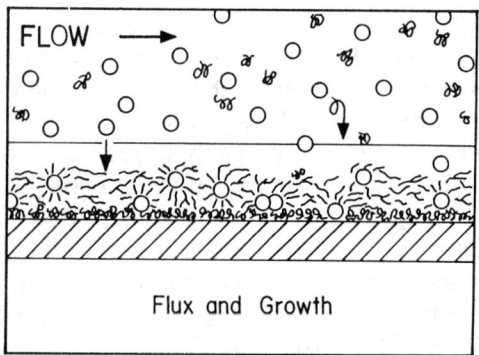

Fig. 11. Continued transport and attachment of microbial cells as well as growth and other metabolic processes within the biofilm.

Stoichiometry

Process rate / Fundamental process		Reactants				Products			
					Biomass		Product		Metabolite
Process	rate	Substrate s	Nutrient z	Electron acceptor e	x_T	x_d	p_e	p_i	a
Growth	μ	−	−	−	+		+	(+)	+
Maintenance									
Exogenous	m	−	−	−			++		++
Endogenous	k_e		+	−	−	(+)	+	−	
Product formation	k_p	−	−	−			+	+	+
Death									
Loss of viability	k_d				−	+			
Lysis	k_L	(+)	(+)		−	(+)	+		
Observed rate		q_s	q_z	q_e	μ_n		q_p		q_a

where q = specific production or removal rate (t^{-1})
μ = specific growth rate or specific biomass production rate (t^{-1})
x_T = total biomass concentration (ML^{-3})
x_d = inert solids concentration (ML^{-3})
p_e = extracellular microbial product concentration (ML^{-3})
p_i = intracellular microbial product concentration (ML^{-3})
s = substrate concentration (ML^{-3})
z = nutrient concentration (ML^{-3})
e = electron acceptor concentration (ML^{-3})
μ_n = net solids production rate (t^{-1})

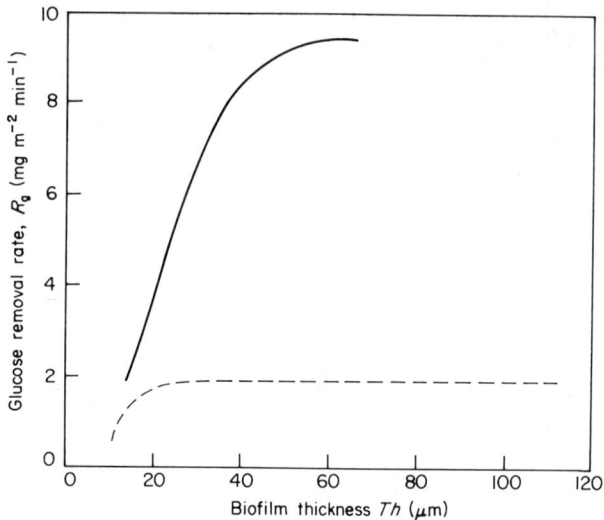

Fig. 12. The influence of biofilm thickness and glucose loading rate on glucose removal rate by a biofilm. Solid curve $R_L = 9.9 \text{ mg m}^{-2} \text{min}^{-1}$. Dashed curve $R_L = 2.0 \text{ mg m}^{-2} \text{min}^{-1}$.

presumably roll along the surface until another surface adhesion site is found.

The nature of the surface is an important factor affecting adhesion. Wettability or critical surface tension, is the property used most frequently to describe surface characteristics in microbial attachment studies (Dexter, 1976; Fletcher and Loeb, 1979). In seawater, cell attachment increased with increasing critical surface tension of the surface (including glass, copper, polyethylene, teflon) with the exception of the copper surface on

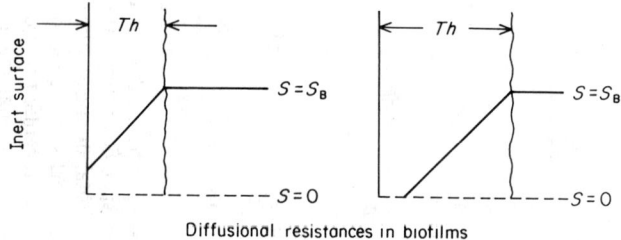

Fig. 13. As biofilm thickness increases beyond the depth of substrate (or oxygen) penetration, substrate (or oxygen) removal rate becomes independent of biofilm thickness.

which fewer cells attached. The copper may inhibit cell attachment by inhibiting a metabolic process necessary for attachment. Even so, there are many examples of biofilm formation on cupronickel condenser surfaces.

The presence of multivalent cations (expecially Ca^{2+}, Mg^{2+} and Fe^{3+}) also influence the attachment process, possibly by altering surface characteristics or by bridging cellular anionic polyelectrolytes to anionic polyelectrolytes adsorbed on the wetted surface.

4.2.4. Metabolism by the Attached Microbial Cells

Restricting our discussion to chemosynthetic organisms, the attached microbial cells assimilate reduced organic or inorganic compounds, nutrients and oxygen or some other electron acceptor. The process yields energy with which the cells reproduce, maintain their internal structure and form extracellular products (Fig. 11). Therefore, *growth, maintenance* and *product formation* are *fundamental processes* carried out by microbial cells in the presence of sufficient nutrients (Table 10). If nutrients are depleted, or toxic substances are present, *death* and/or *lysis* may occur.

The rates of the fundamental microbial processes are difficult to measure. The *observed rates* (last row, Table 10) are usually rate of substrate (the growth-limiting nutrient) removal, electron acceptor (usually oxygen) removal, biomass production or product formation.

The stoichiometry of each fundamental process can be measured in certain laboratory systems (e.g. chemostats). The rows in Table 10 qualitatively represent the stoichiometry of each fundamental process (− refers to reactant and + refers to product).

Analysis of rate and stoichiometry of processes within a biofilm are frequently complicated by significant mass transfer resistances in the liquid or diffusional resistances within the biofilm.

Trulear and Characklis (1982) have observed substrate removal rate in an experimental biofilm reactor. The substrate removal rate increases in proportion to biofilm thickness up to a critical thickness beyond which removal rate remains constant (Fig. 12). The critical, or "active", thickness is observed to increase with substrate concentration. This behaviour is confirmed by other investigators (LaMotta, 1976; Kornegay and Andrews, 1968) and is attributed to nutrient diffusional limitations within the biofilm. Once the biofilm thickness exceeds the depth of substrate (or oxygen) penetration into the biofilm (Fig. 13), the removal rate is unaffected by further biofilm accumulation.

Observed substrate removal rate cannot be used to distinguish between growth, maintenance, product formation and death. It seems clear from other data (Bryers, 1979) that product formation (primarily polysaccharide)

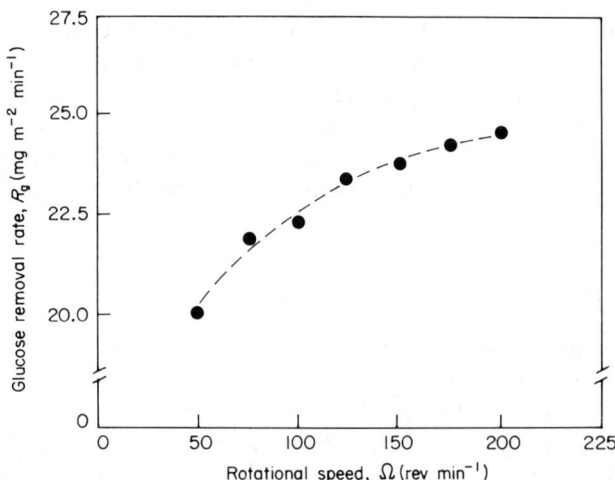

Fig. 14. Influence of rotational speed on glucose removal rate by a biofilm in an annular reactor. $Th = 112$ μm. $R_L = 27.0$ mg m^{-2} min^{-1}.

is significant in the early stages of biofilm formation. Maintenance requirements or biomass decay become important as the biofilm gets thicker and substrate does not entirely penetrate the biofilm. These other process rates have not been measured and are critical for determining stoichiometric coefficients and predicting biofilm development rates.

The substrate removal rate is also dependent on fluid velocity (Fig. 14). At low fluid velocities, a relatively thick mass transfer boundary layer (δ_m) can cause a liquid phase diffusional resistance which decreases substrate concentration at the liquid–biofilm interface and thereby decreases substrate removal rate (Fig. 15).

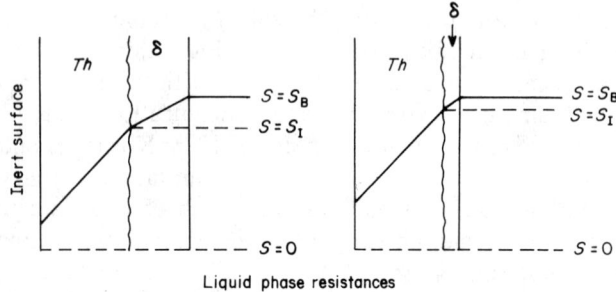

Fig. 15. The mass transfer boundary layer (δ_m) decreases with increasing fluid velocity past the biofilm interface resulting in a higher effective substrate concentration at the biofilm–fluid interface (S_I).

6. PROCESS ANALYSIS IN MICROBIAL SYSTEMS 219

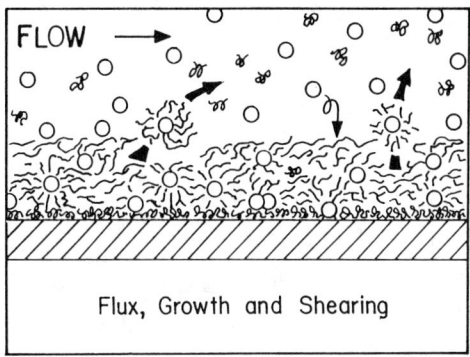

Fig. 16. Transport, attachment and growth increase the accumulated mass of the biofilm while detachment processes decrease the attached mass.

4.2.5. *Detachment of Biofilm*

As the biofilm grows thicker, the fluid shear stress at the biofilm interface generally increases. Also as biofilms grow thicker, the potential for substrate, oxygen or nutrient limitation in the deeper portions is great. These limitations may weaken the biofilm matrix and cause detachment (Fig. 16). Trulear and Characklis (1982) report that the biofilm detachment rate increases with increasing biofilm mass (Fig. 17) and that detachment rate increases with fluid shear stress (Fig. 18).

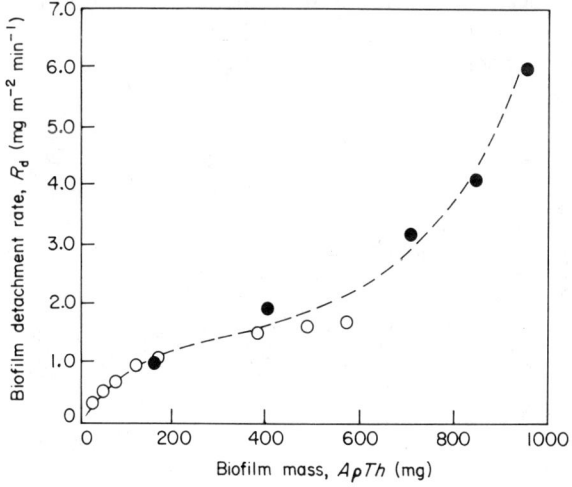

Fig. 17. Influence of biofilm mass on biofilm detachment rate at a constant fluid shear stress. ●, $R_L = 37.2$ mg m^{-2} min^{-1}; ○, $R_L = 4.2$ mg m^{-2} min^{-1}.

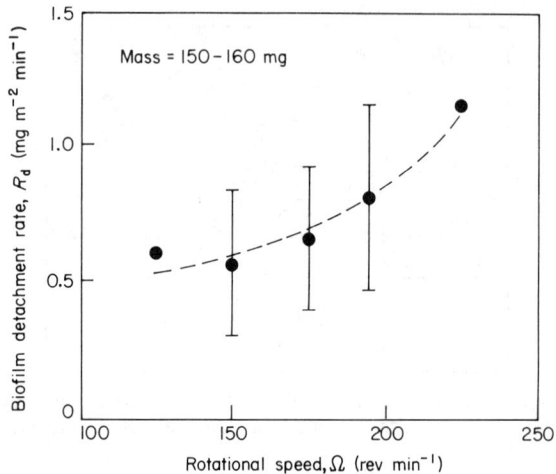

Fig. 18. Influence of fluid shear stress on biofilm detachment rate at a constant attached biomass.

Techniques for determining strength of adhesion and strength of deposit are necessary to further understanding of the detachment process.

4.2.6. Overall Rate of Biofilm Development

A general mathematical model for microbial processes in a continuous stirred tank reactor (CSTR), based on material balances, is presented in Table 11. The model considers microbial activity in the bulk fluid as well as the reactor surfaces.

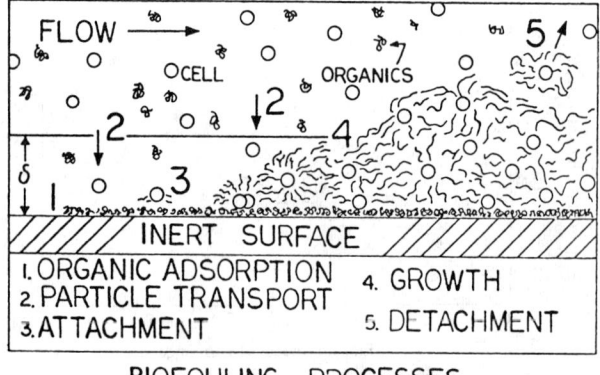

Fig. 19. A summary diagram of processes contributing to biofilm accumulation.

Table 11. Mathematical representation of microbial processes, including biofilm formation, occurring in a CSTR (see Fig. 19).

Substrate material balance

$$V\frac{ds}{dt} = F(s_1 - s) - NA - \frac{(\mu - m)xV}{Y_x} \quad (37)$$

Suspended biomass material balance

$$V\frac{dx}{dt} = F(x_i - x) + \mu xV - R_A A + R_D A \quad (38)$$

Accumulation of biofilm

$$\frac{dB}{dt} = NAY_B - R_D A + R_A A - R_E B \quad (39)$$

Accumulation of total reactor biomass

$$\frac{dM_t}{dt} = V\frac{dx}{dt} + \frac{dB}{dt} = F(x_i - x) + \mu xV + NAY_B - R_E B \quad (40)$$

where A = wetted surface area (L^2)
  B = biofilm mass (M_B)
  F = volumetric flow rate $(L^3 t^{-1})$
  m = maintenance coefficient (t^{-1})
  M_t = total reactor biomass (M)
  N = substrate flux into the biofilm $(M_S L^{-2} t^{-1})$
  R_A = rate of suspended biomass adsorption onto the biofilm $(M_x L^{-2} t^{-1})$
  R_D = rate of biofilm detachment $(M_B L^{-2} t^{-1})$
  R_E = rate of biofilm decay (e.g., lysis, endogenous respiration) (t^{-1})
  $s(s_i)$ = (input) substrate concentration $(M_s L^{-3})$
  t = time (t)
  $x(x_i)$ = (input) suspended biomass concentration $(M_x L^{-3})$
  Y_B = yield coefficient for biofilm $(M_B M_s^{-1})$
  Y_x = yield coefficient for suspended biomass $(M_x M_s^{-1})$
  V = reactor volume (L^3)
  μ = specific growth rate of suspended biomass (t^{-1})

Biofilm development is the net result of several processes occurring in series and parallel (Fig. 19). The development of a biofilm is adequately described by a sigmoidal-shaped curve (Fig. 20). The slope of this curve at a particular time is the *net* biofilm *development rate* and is also plotted in Fig. 20. The rate increases to a maximum value corresponding to the sigmoidal inflection and then decreases to zero. Net biofilm development rate is expressed as follows (Table 11):

$$\frac{dB}{dt} = NAY_B - R_D A + R_A A - R_E B \quad (39)$$

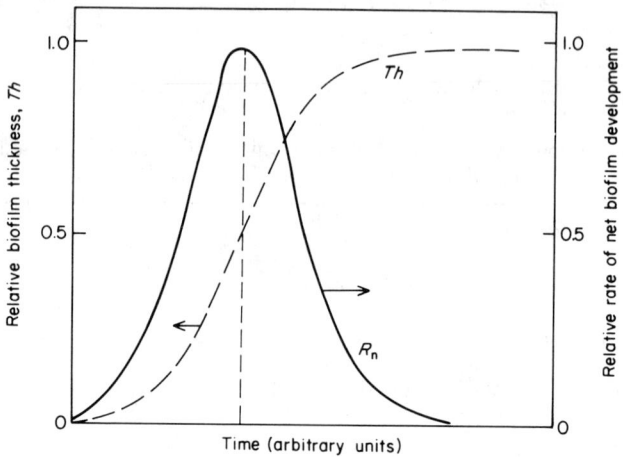

Fig. 20. Progression of net biofilm development is described by a sigmoidal-shaped curve. Net biofilm development rate is the slope of the sigmoidal-shaped curve at any time.

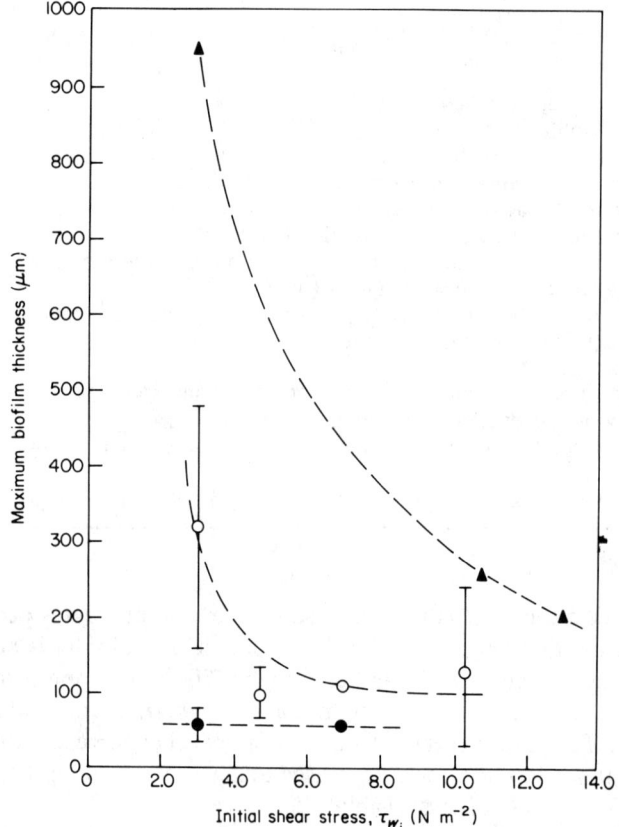

Fig. 21. Influence of fluid shear stress and substrate loading on plateau (or maximum) biofilm thickness. $R_L = 18.30$ mg m^{-2} min^{-1}. $R_L = 2.6$–7.0 mg m^{-2} min^{-1}. $R_L = 1$–15 mg m^{-2} min^{-1}.

where NAY_B is the attached biomass production rate, $\dot{R}_D A$ is the biofilm detachment rate, $R_A A$ is the adsorption rate of cells and $R_E B$ is the endogenous respiration rate. At steady state, thickness remains constant.

The effect of fluid velocity on the plateau (or steady state) biofilm thickness is illustrated in Fig. 21 for various substrate loadings. At high substrate loadings, increasing fluid velocity increases biofilm detachment rate which minimizes the plateau biofilm thickness. However, at low substrate loadings, fluid velocity seems to have no measurable effect on the plateau thickness. Trulear and Characklis (1982) have demonstrated that plateau biofilm mass exhibits a maximum when fluid velocity is increased. At low fluid velocities, mass transfer limits the rate of biofilm production. Therefore, increasing fluid velocity increases substrate flux into the biofilm and net biofilm development rate increases. As fluid velocity continues to increase, biofilm detachment rate becomes the dominant process and net biofilm development begins to decrease.

4.3 Effects of Biofilms on Fluid Frictional Resistance

Increase in fluid frictional resistance due to biofilm accumulation when flow rate is maintained constant causes an increase in pressure drop and power requirements for pumping as shown in Fig. 22 (Picologlou et al.,

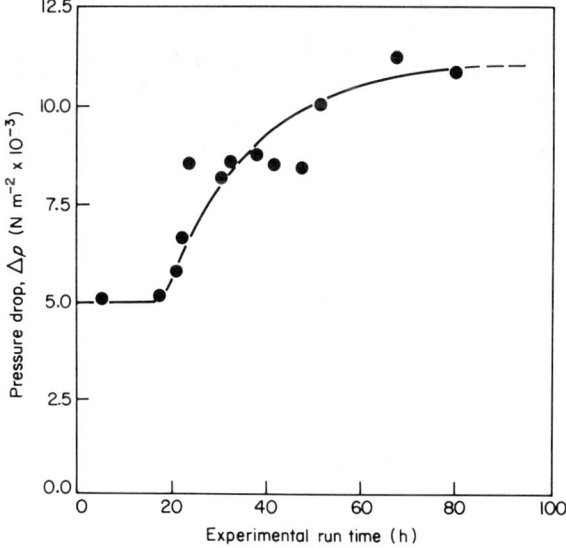

Fig. 22. Change in pressure drop with time due to biofilm formation. Experiment was conducted at constant fluid velocity. Initial $\tau_w = 6.8\,\text{N m}^{-2}$. $\bar{v} = 150\,\text{cm s}^{-1}$.

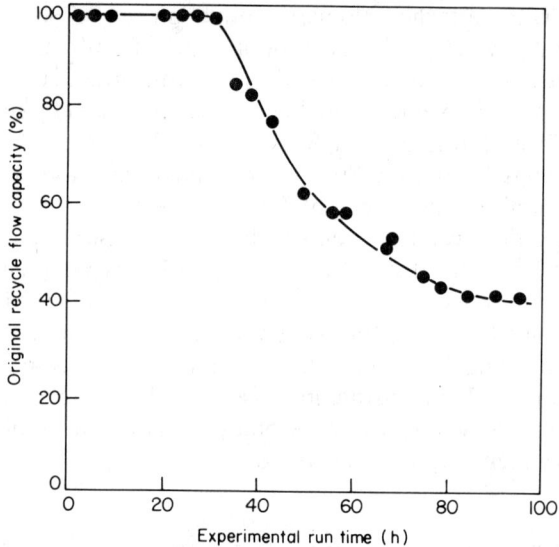

Fig. 23. Change in volumetric flow rate with time due to biofilm formation. Experiment was conducted at constant pressure drop. $\tau_w = 7.9\,\text{N m}^{-2}$. Initial $\bar{v} = 185\,\text{cm s}^{-1}$.

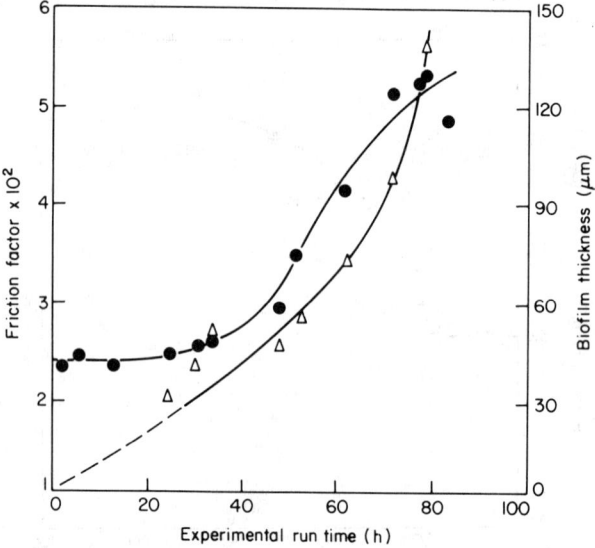

Fig. 24. Change in friction factor and biofilm thickness with time for an experiment conducted at constant pressure drop. $\tau_w = 7.9\,\text{N m}^{-2}$. Initial $\bar{v} = 180\,\text{cm s}^{-1}$. ●, friction factor; △, biofilm thickness.

1980). Conversely, if pressure drop is held constant, flow capacity is reduced. Figure 23 indicates that flow capacity was reduced to 42% of the original capacity in a 100-h laboratory experiment (Picologlou et al., 1980).

Frictional resistance can be represented by a dimensionless friction factor given by the following equation:

$$f = 2.0 \frac{d}{L} \frac{\Delta_p}{\rho \bar{v}^2} \quad (41)$$

where f = friction factor (dimensionless)
d = tube diameter [L]
ρ = fluid density [ML^{-3}]
$\bar{v}$ = average fluid velocity [Lt^{-1}]
Δp = pressure drop along length L [ML^{-1}t^{-2}]
L = length between pressure ports [L]

The change in friction factor and biofilm thickness with time is shown in Fig. 24 for a laboratory tubular reactor. Dehart (1979) has observed similar behaviour in a tubular reactor in the field (Fig. 25).

The friction factor is related to the Reynolds number and the equivalent sand roughness k_s through the empirical Colebrook–White relation. This equation correlates friction factor to the Reynolds number for various "commercially rough" pipes throughout the hydraulically smooth, transition and fully rough regimes. The Colebrook–White equation, solved for the equivalent sand roughness k_s yields

$$k_s = \frac{d}{2} 10^{(0.87 - 0.50f^{-1/2})} - \frac{18.70}{\mathrm{Re}\, f^{1/2}} \quad (42)$$

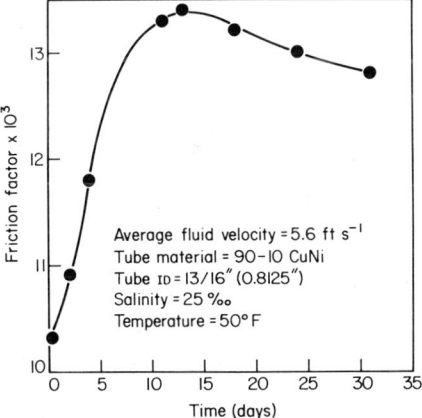

Fig. 25. Change in pressure drop due to biofilm formation at a field location (Dehart, 1979).

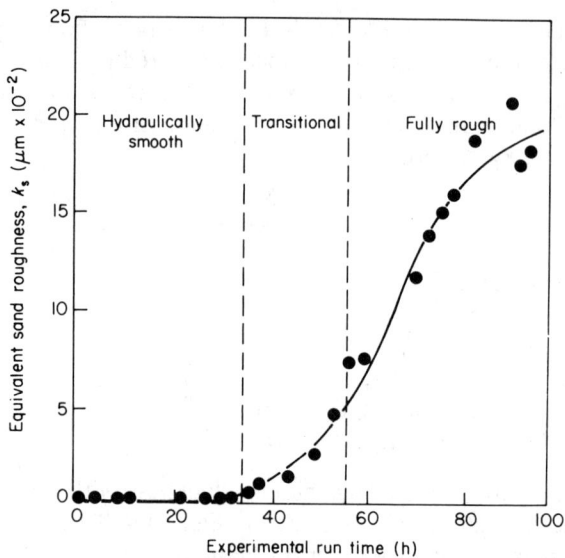

Fig. 26. Change in equivalent sand roughness with time due to biofilm formation. Experiment was conducted at constant pressure drop. $\tau_w = 7.9\,\text{N m}^{-2}$. Initial $\bar{v} = 163\text{--}185\,\text{cm s}^{-1}$.

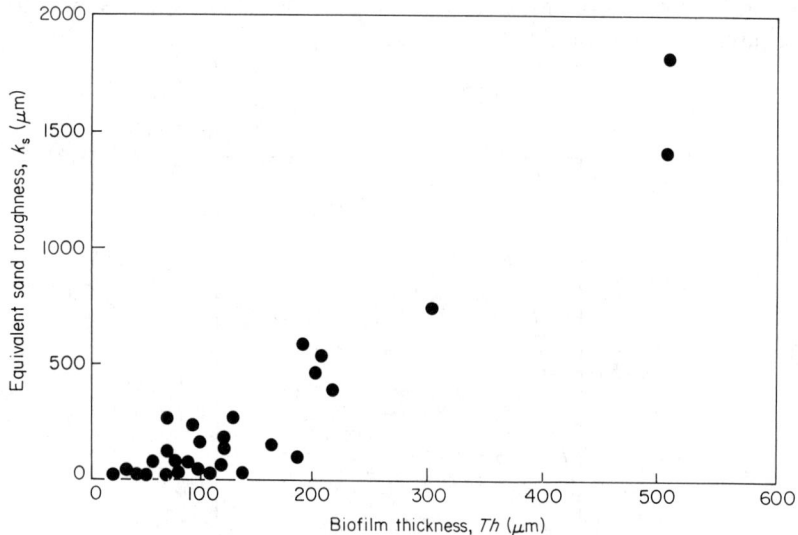

Fig. 27. Change in calculated equivalent sand roughness with biofilm thickness for several experiments conducted at constant pressure drop. $\tau_w = 6.5\text{--}7.9\,\text{N m}^{-2}$. Initial $\bar{v} = 163\text{--}185\,\text{cm s}^{-1}$.

6. PROCESS ANALYSIS IN MICROBIAL SYSTEMS

where d = tube diameter [L]
$\mathrm{Re} = \bar{v}d/\nu$ = Reynolds number (dimensionless)
ν = kinematic viscosity [$L^2 t^{-1}$]
This expression can be used to compute an equivalent sand roughness for the biofilm from a measurement of the flow rate and pressure drop. Figure 26 indicates the progression of k_s with time and Fig. 27 presents the change in k_s with biofilm thickness for the range of shear stress investigated by Picologlou et al. (1980).

Determination of the flow regime (smooth, transitional or fully rough) depends on the magnitude of k_s relative to the size of the viscous sublayer (δ_1):

$$\delta_1 = \frac{10d}{\mathrm{Re}} \left(\frac{f}{2}\right)^{-0.5} \quad (43)$$

More specifically, when $k_s < \delta_1$, the pipe is considered hydraulically smooth; when $14\delta_1 > k_s > \delta_1$ the flow is in the transitional regime; when $k_s > 14\delta_1$ the flow is in the fully rough regime (Schlichting, 1968).

Frictional resistance of biofilms grown under constant pressure drop (i.e. constant shear stress) have been compared to the frictional resistance of pipes with a rigid roughness as given by the Colebrook–White equation. The following was observed:

(1) Frictional resistance due to biofilms shows a similar dependency on Reynolds number as frictional resistance due to commercially rough pipe surface.
(2) Frictional resistance is dependent on biofilm thickness.
(3) Frictional resistance does not increase above the hydraulically smooth pipe value until a critical biofilm thickness is attained.

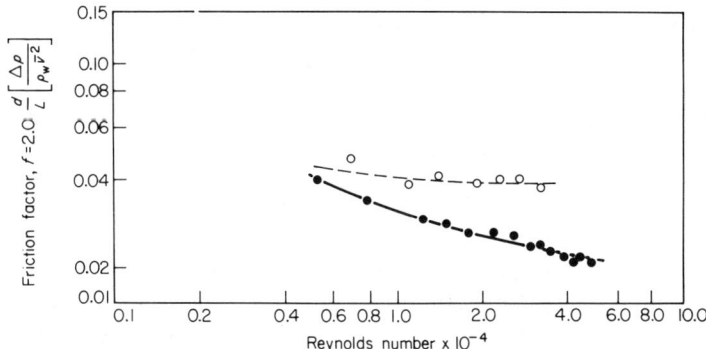

Fig. 28. Change in friction factor as a function of Reynolds number and roughness due to biofilm formation. ○, $k_s/d = 0.008$; $Th = 320\,\mu\mathrm{m}$. ● clean tube.

The Blasius–Stanton or Moody diagram (Moody, 1944) can be used to compare frictional resistance due to biofilm with frictional resistance of rigid rough surfaces. The Blasius–Stanton diagram is a plot of friction factor versus Reynolds number for a series of pipes with different equivalent sand roughness; the friction factor in a pipe with a rigid rough surface depends on both the relative roughness and the Reynolds number.

The relationship between friction factor and Reynolds number for a fouled circular tube is presented in Fig. 28. The friction factors and Reynolds numbers presented have not been corrected for the pipe constriction resulting from the biofilm. This figure shows the dependency of friction factor on Reynolds number is the same as for a tube with a rigid rough surface within the range of Reynolds number investigated (5000–48 000). This data was obtained by reducing, in steps, the shear stress from its initial value in a given experiment and calculating friction factor and Reynolds number at each step. The shear stress was reduced from the initial condition to minimize detachment of biofilm during the experiment.

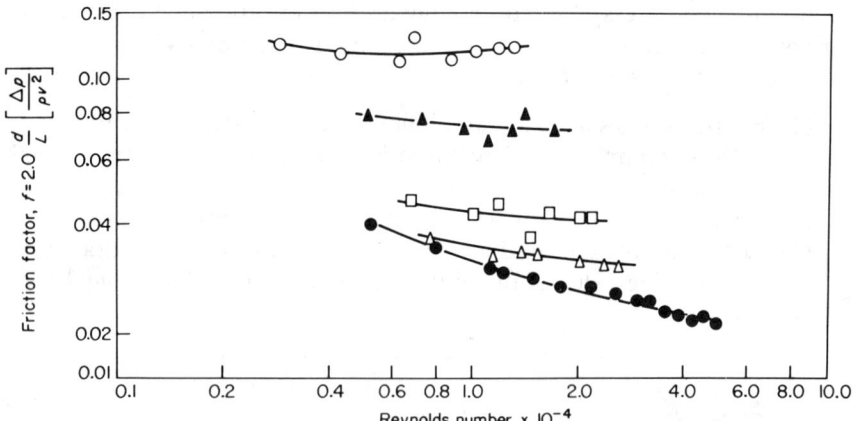

Fig. 29. Change in friction factor as function of Reynolds number and roughness at different stages of biofilm development. ○, $k_s/d = 0.157$; $Th = 500\,\mu m$. ▲, $k_s/d = 0.062$; $Th = 300\,\mu m$. □, $k_s/d = 0.014$; $Th = 165\,\mu m$. △, $k_s/d = 0.002$; $Th = 40\,\mu m$. ●, clean tube.

Figure 29 indicates the relationship between friction factor and Reynolds number within a single experiment at different stages of biofilm development; friction factor increases with biofilm thickness. The relationship between biofilm thickness and friction factor at a wall shear stress from 6.5–7.9 N m^{-2} is shown in Fig. 30. Friction factor is dependent on biofilm

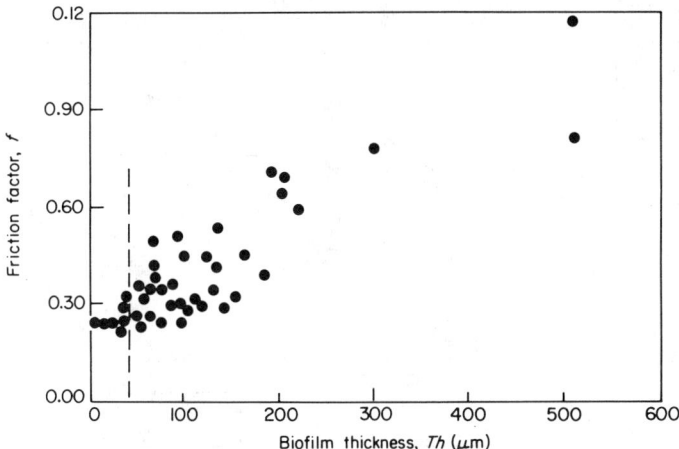

Fig. 30. Change in friction with biofilm thickness at constant pressure drop indicating viscous sublayer thickness. $\tau_w = 6.5$–7.9 N m^{-2}. Initial $\bar{v} = 163$–185 cm s^{-1}.

thickness after a critical thickness (Th_c) approximately equal to the thickness of the viscous sublayer (δ_1) is attained.

Conceptually, Th_c corresponds to the stage of biofilm development at which surface irregularities protrude through the viscous sublayer. Until this stage, the biofilm lies completely within the viscous sublayer ($k_s < \delta_1$) and friction factor does not increase (the tube is hydraulically smooth). For a wall shear stress of 6.5–7.9 N m^{-2}, the viscous sublayer is approximately equal to 40 μm; this compares well with the observed $Th_c = 30$–35 μm for the same wall shear stress range.

Although the frictional resistance effects of biofilm can be adequately described by formulae and concepts suitable for rigid rough surfaces, the conclusion should not be drawn that indeed the biofilm presents a rigid rough surface to the flow. Such a notion is an oversimplification and cannot account for all experimental observations (Sherwood *et al.*, 1975).

Finally, frictional resistance measurements provide a relatively simple method for determining liquid mass transfer resistance in some biofilm systems since frictional resistance and liquid mass transfer resistance are related (Sherwood *et al.*, 1975).

4.4. Effects of Biofilms on Heat-Transfer Resistance

Biofilm development and resulting fluid frictional resistance have been discussed and both influence heat transfer. Changes in heat-transfer resistance arise from the combined effects of increased biofilm thickness (con-

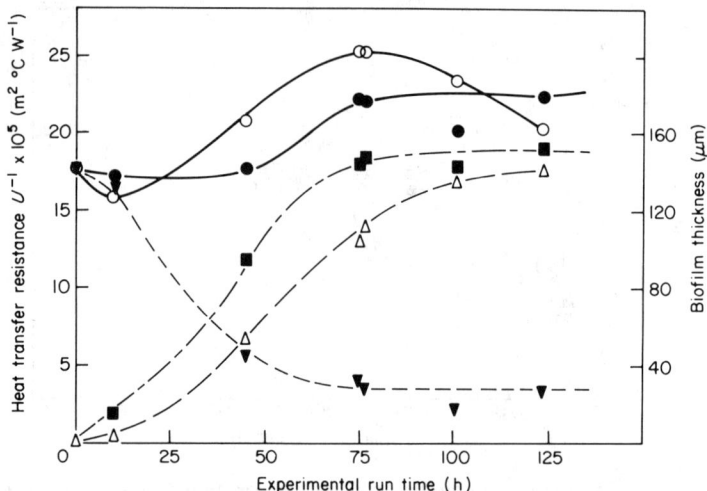

Fig. 31. Changes in convective and conductive heat transfer resistance as a result of biofilm development. ○, measured overall resistance. ●, total resistance. ■, conductive resistance. △, biofilm thickness. ▲, convective resistance.

ductive heat transfer) and increased frictional resistance (convective heat transfer).

Conductive heat transfer can be related to biofilm thickness and its effective thermal conductivity. Experimental biofilm thermal conductivity determinations indicate no significant difference from that of water at the

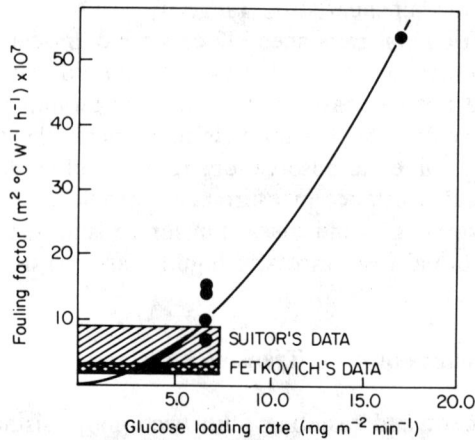

Fig. 32. Influence of substrate loading rate on heat transfer fouling reactor. The ranges measured by others are indicated for comparison (see Table 12).

6. PROCESS ANALYSIS IN MICROBIAL SYSTEMS   231

same temperature (see Table 6). This is not surprising since biofilm is approximately 98–99% water.

Convective heat transfer results from fluid mixing or motion, and can be related to momentum transfer or frictional resistance. Colburn (1933) correlated convective heat transfer in tubes to friction factor and properties of the fluid. The Colburn relationship is only useful when the biofilm is thicker than the viscous sublayer.

Overall heat-transfer resistance due to biofouling film development can then be calculated if the following are known:

(1) Biofilm thickness and biofilm thermal conductivity.
(2) Frictional resistance.
(3) Wall temperature and bulk temperature.

Figure 31 describes a typical experiment (Characklis *et al.*, 1981) in a tubular reactor and illustrates the relative effects of conductive and convective heat-transfer resistance on overall heat-transfer resistance.

Heat-transfer resistance was consistently observed to decrease upon initial exposure to the fouling fluid. Presumably, a very thin biofilm causes a decrease in convective heat-transfer resistance which is larger than the resulting increase in conductive heat-transfer resistance.

The rate of change in heat-transfer resistance is strongly dependent on substrate loading. R_f is a fouling factor traditionally used by engineers to quantify the influence of biofilms (and other deposits) on heat-transfer rate. Rate of change in R_f measured in natural seawater systems are compared to laboratory freshwater data in Fig. 32. Table 12 describes the experimental conditions in which the data were obtained. Neither carbon nor nutrient concentrations were determined in the latter two studies. However, soluble organic carbon concentrations are estimated at between 0.5 and 10 mg l^{-1}.

Kirkpatrick *et al.* (1980) have modelled the heat and mass transfer occurring in a heat exchange tube as a biofilm develops. In a typical heat exchanger, results indicate a significant decrease in heat transfer. For

Table 12. Description of experimental systems for R_f measurements reported in Fig. 32.

	Characklis *et al.*, 1981	Ritter *et al.*, 1977	Fetkovich *et al.*, 1978
Surface	Al 6061-T6	Titanium	Cupronickel
Surface temperature (°C)	39–45	26–38	21
Fluid velocity (cm s^{-1})	81	60–120	90–180

systems of interest, the biofilm is relatively uniform over the length of the heat exchange tube. In tubes with combined heat and mass transfer, the biofilm thickness varies appreciably with fluid temperature. The assumed relationships between temperature and biofilm development rates in their model have been partially verified by Stathopoulos (1981).

5. Summary

The basis and methods for a process analysis of biofilm development have been presented. The purposes were as follows:

(1) To familiarize the reader with process analysis based on conservation and constitutive principles.
(2) To present the assumptions frequently made in such an analysis.
(3) To present a framework for analysis of the rate of biofilm development, extent of biofilm development and influence of biofilms on energy losses.

Biofilms are emerging as a most critical factor affecting natural aquatic systems, water distribution systems, wastewater treatment systems, heat exchangers, shipping and human disease. More attention is being directed to this behaviour.

Acknowledgements

The author gratefully acknowledges partial financial support from the following during preparation of the manuscript: Office of Naval Research, National Science Foundation and Calgon Corporation. May Mace and Sharlene Vehnekamp typed the manuscript.

References

Anderson, M. R., Vacarro, R. F. and Toner, R. C. (1977). "Biofouling Control Procedures" (L. D. Jensen, ed.). Marcel Dekker, New York.
Arnold, G. E. (1936). *Eng. News Rec.* **116**, 774–775.
Atomic Energy Commission (1955). "Reactor Handbook", Vol. 2, AECD-3646. U.S.G.P.O., Washington, D.C.
Baillod, R. D. and Boyle, W. C. (1970). *J. Sanit. Eng. Div., A.S.C.E.* **96**(SA4), 525–545.
Beal, S. K. (1970). *Nucl. Sci. Eng.* **40**, 1–11.
Brash, J. L. and Samak, Q. M. (1979). *J. Colloid, Interface Sci.* in press.
Browne, L. W. B. (1974). *Atmos. Environ.* **8**, 801.
Bryers, J. D. (1979). Ph.D. Thesis, Rice University, Houston, Texas.

Bryers, J. D. and Characklis, W. G. (1979). In "Condenser Biofouling Control" (J. F. Garey et al., eds.), pp. 169–183. Ann Arbor Science, Ann Arbor, Michigan.
Busch, A. W. (1971). "Aerobic Biological Treatment of Wastewaters." Oligodynamics Press, Houston, Texas.
Characklis, W. G. (1973). Water Res. 7, 1113–1127.
Characklis, W. G. (1980). "Biofilm Development and Destruction." Final Report, Electric Power Research Inst., RP 902-1, Palo Alto, California.
Characklis, W. G., Nimmons, M. J. and Picologlou, B. F. (1981). Heat Transfer Eng., 3, 23–37.
Churchill, S. W. (1974). "The Interpretation and Use of Rate Data: The Rate Concept." McGraw-Hill, New York.
Cleaver, J. W. and Yates, B. (1975). Chem. Eng. Sci. 30, 983.
Cleaver, J. W. and Yates, B. (1976). Chem. Eng. Sci. 31, 147.
Colburn, A. P. (1933). Trans. AIChE 29, 174.
Corpe. W. A. (1970) Develop. Ind. Microbiol. 11, 402–412.
Corpe, W. A. (1978). In "Microbiology of Power Plant Thermal Effluents" (R. M. Gerhold, ed.), pp. 57–66. University of Iowa, Iowa City, Iowa.
Costerton, J. W., Geesey, G. G. and Cheng, K. J. (1978). Sci. Amer. 238, 86–95.
Dehart, R. (1979). Marine Research, Inc., Sandwich, Massachusetts. Personal Communication.
DePalma, V. A., Goupil, D. W. and Akers, C. K. (1979). Proc. 6th OTEC Conf. (June 1979).
Dexter, S. C. (1976). Proc. 4th Int. Cong. Mar. Corros. Fouling, Juan-Les-Pins, Antibes, France.
Fetkovich, J. G., Granneman, G. N., Mahalingam, L. M. and Meier, D. L. (1978). Proc. OTEC Biofouling Corros. Symp. (Oct. 1977), pp. 237–380.
Fletcher, M. (1977). Can. J. Microbiol. 23, 1–6.
Fletcher, M. and Loeb, G. I. (1979). Appl. Environ. Microbiol. 37, 67–72.
Frederickson, A. G., Ramkrishna, D. and Tsuchiya, H. M. (1967). Math. Biosa. 1, 327–374.
Frederickson, A. G., Megee, R. D. and Tsuchiya, H. M. (1970). Adv. Appl. Microbiol. 13, 419–465.
Friedlander, S. K. and Johnstone, H. F. (1957). Ind. Eng. Chem. 49, 1151–1156.
Gunsalus, I. C. and Stanier, R. Y. (1960). "The Bacteria," Vol. 1. Academic Press, New York.
Herbert, D. (1961). Symp. Soc. Gen. Microbiol. 11, 391–416.
Himmelblau, D. M. and Bischoff, K. B. (1968). "Process Analysis and Simulation." Wiley, New York.
Jenkins, D. (1980). Unpublished results.
Jones, H. C., Roth, I. L. and Sanders, W. M. (1969). J. Bacteriol. 99, 316–325.
Kirkpatrick, J. P., McIntire, L. V. and Characklis, W. G. (1980). Water. Res. 14, 117–127.
Kornegay, B. H. and Andrews, J. F. (1967). J. Water Pollut. Contr. Fed. 40, R460–R468.
LaMotta, E. J. (1976). Environ. Sci. Technol. 10, 765–769.
Levenspiel, O. (1972). "Chemical Reaction Engineering." Wiley, New York.
Lin, C. S., Moulton, R. W. and Putnam, G. L. (1953). Ind. Eng. Chem. 45, 636–640.
Lister, D. H. (1981). In "Fouling of Heat Transfer Equipment" (E. F. C.

Somerscales and J. G. Knudsen, eds.), pp. 135–200. Hemisphere, Washington, D.C.
Loeb, G. I. and Neihof, R. A. (1975). *In* "Applied Chemistry at Protein Interfaces," pp. 319–335. Advances in Chemistry Series, Vol. 145. A.C.S., Washington, D.C.
Marshall, K. C. (1976). "Interfaces in Microbial Ecology." Harvard University Press, Cambridge, Massachusetts.
Marshall, K. C., Stout, R. and Mitchell, R. (1971). *J. Gen. Microbiol.* **68**, 337–348.
Matson, J. V. (1975). Ph.D. Thesis, Rice University, Houston, Texas.
Matson, J. V. and Characklis, W. G. (1976). *Water Res.* **10**, 877–885.
Michaelis, L. and Manton, M. L. (1913). *Zeitschrift für Biochemistry* **49**, 333–369.
Minkus, A. J. (1954). *J. New Engl. Water Workers Ass.* **68**, 1–10.
Moody, L. F. (1944). *Trans. Amer. Soc. Mech. Eng.* **66**, 671.
Mueller, J. A., *et al.* (1966). Proc. 21st Ann. Purdue Ind. Waste Conf., pp. 962–995.
Perry, R. H. and Chilton, C. H. (eds.) (1973). "Chemical Engineers Handbook." McGraw-Hill, New York.
Picologlou, B. F., Zelver, N. and Characklis, W. G. (1980). *J. Hydraul. Div., A.S.C.E.* **106**, 733–746.
Pipes, D. M. (1974). M.Sc. Thesis, Rice University, Houston, Texas.
Pollard, A. L. and House, H. E. (1959). *J. Power Div., A.S.C.E.* **85**, 163–171.
Ritter, R. B., Suitor, J. W. and Cypher, G. A. (1977). "Thermal Fouling Rates of 90–10 Copper-Nickel and Titanium in Seawater Service". Report to INCRA, HTRI, Alhambra, California.
Roels, J. A. and Kossen, N. W. F. (1978). *Progr. Ind. Microbiol.* **14**, 95–203.
Rouhiainen, P. O. and Stachiewicz, J. W. (1970). *J. Heat Transfer (Trans. A.S.M.E.)* **92**, 169.
Schaechter, M., Maaløe, O. and Kjeldgaard, N. O. (1958). *J. Gen. Microbiol* **19**, 592–606.
Schlichting, H. (1968). "Boundary Layer Theory." McGraw-Hill, New York.
Sherwood, T. K., Pigford, R. L. and Wilkie, C. R. (1975). "Mass Transfer." McGraw-Hill, New York.
Stathopoulos, N. A. (1981). M.Sc. Thesis, Rice University, Houston, Texas.
Tomlinson, T. G. and Snaddon, D. M. (1966). Proc. 21st Ann. Purdue Ind. Waste Conf., pp. 962–995.
Trulear, M. G. (1983). Ph.D. Thesis, Montana State University, Bozeman, Montana.
Trulear, M. G. and Characklis, W. G. (1982). *J. Water Pollut. Contr. Fed.*, **54**, 1288–1301.
Weast, R. D. (ed.) (1973). "Handbook of Chemistry and Physics." CRC Press, Cleveland, Ohio.
Wells, A. C. and Chamberlain, A. C. (1967). *Brit. J. Appl. Phys.* **18**, 1793.
Williams, F. M. (1967). *J. Theoret. Biol.* **15**, 190–207.
Williamson, K. J. and McCarty, P. L. (1976). *J. Water Pollut. Contr. Fed.* **48**, 281–296.
Zelver, N. (1979). M.Sc. Thesis, Rice University, Houston, Texas.
Zobell, C. E. (1943). *J. Bacteriol.* **46**, 39–59.

Chapter 7

Theory and Practice of Time-Domain Techniques

J. F. Dalrymple

Department of Management Science and Technology Studies, University of Stirling, Stirling, U.K.

J. M. Crowther

Department of Applied Physics, University of Strathclyde, Glasgow, U.K.

In biology several variables may be measured, all of which are assumed to be related to a single independent variable, time. Such data may result from an experiment designed to test a specific hypothesis. Frequently, especially in large-scale industrial applications and in effluent purification processes, it is not practical to perform experiments of this sort. In such cases the time-dependent observations may be analysed by time domain techniques, the subject of this chapter, in order to predict the future behaviour of the system under study and to gain some understanding of how it operates—*Editorial note.*

1. Introduction

The study of biological and environmental systems is made difficult by their natural variability and inherent complexity. It is indeed often difficult both conceptually and practically to isolate a system for study. But assuming that this can be done it is then of interest to know how the system variables will evolve in time. One may distinguish between a passive approach in which one observes the system for a period of time under natural conditions and stimuli, or an active approach in which artificial conditions or stimuli are applied. Whichever approach is adopted, the observer will be faced with the problem of obtaining and analysing sequences of measurements. A sequence of such measurements ordered in time is called a "time series", and the task of time series analysis is to extract as much information as possible concerning the system and its responses.

Some of the measured variables may be internal to the system, others may be external and imposed on the system; some may be continuous (e.g. temperature), others discrete (e.g. daily rainfall totals).

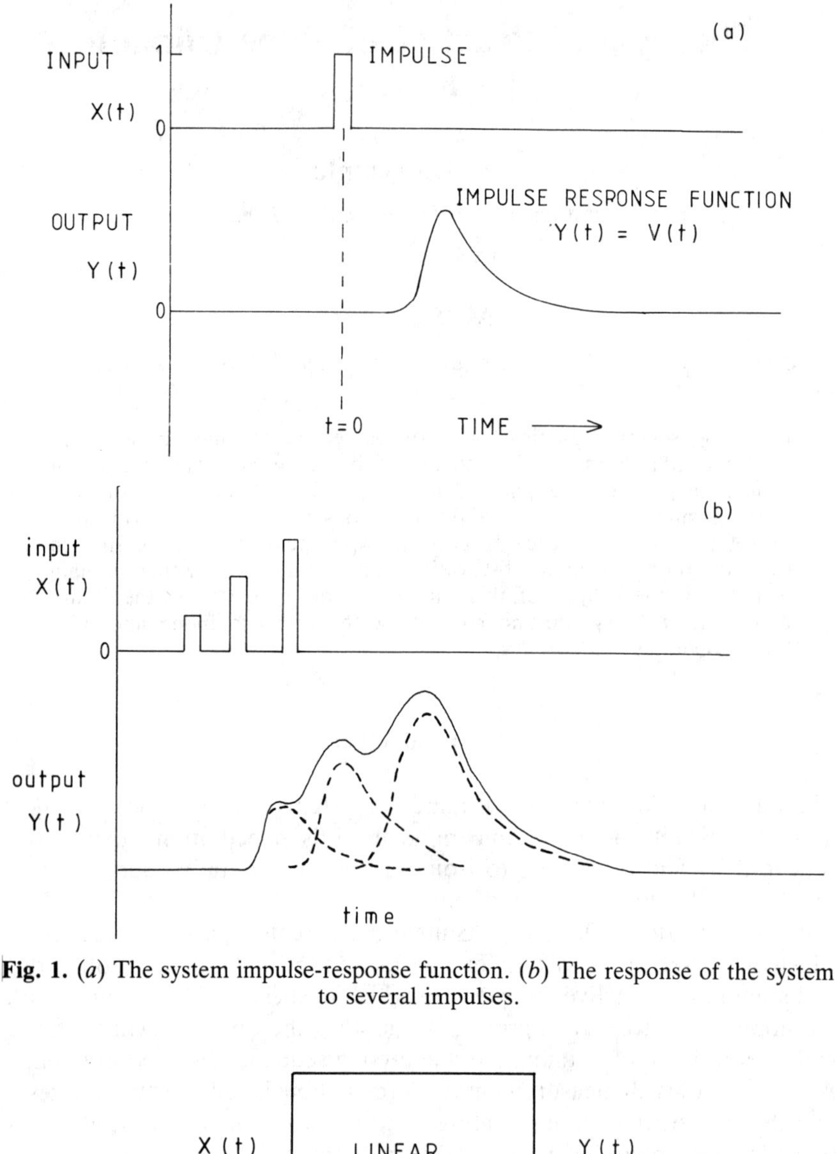

Fig. 1. (*a*) The system impulse-response function. (*b*) The response of the system to several impulses.

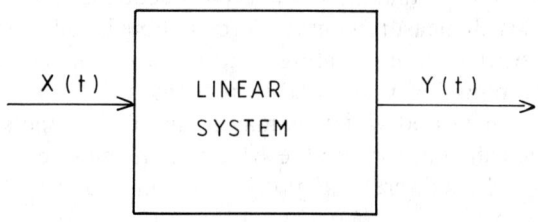

Fig. 2. A linear system with a single input and a single output.

7. THEORY AND PRACTICE OF TIME-DOMAIN TECHNIQUES

There may indeed be variables which, owing to lack of available techniques or resources, cannot be measured yet affect the system significantly. It must also be recognized that the measurements which are made will be subject to measurement error.

The main body of literature of time-series analysis deals with observations which are equally spaced in time, separated by some convenient and appropriate time interval. The exact choice of the time interval depends on the time constants of the system under investigation, and a "rule of thumb" to guide the experiment designer is that the time interval between samples should be long enough to allow the system to have undergone a measurable change between successive samples, but short enough to allow several observations to be made in the course of any significant time constant of the system. Thus, some knowledge of the time development and time constants of the system is a prerequisite for experiments designed to collect data for time-series analysis. Success in time-series analysis is obviously dependent on having a set of good quality data collected with an appropriate frequency at equal intervals for an adequate duration.

Once such a data set has been assembled, there are two possible approaches which can be adopted for time-series analysis. These are frequency-domain analysis and time-domain analysis. For linear systems in the limit of infinite series, the two techniques yield identical results since they are formally equivalent. However, for finite series the results may differ and the choice of approach then depends on the judgement of the individual worker.

Frequency-domain techniques were developed for use in the field of statistical communication theory and electrical engineering, and are well documented by Bendat and Piersol (1966) and Jenkins and Watts (1968). The methods have been adopted in acoustics and other areas where there is reason to believe that the system under consideration is frequency responsive, and where electrical engineering concepts of gain and phase are useful and easily interpretable.

There are, on the other hand, systems which are not intrinsically "frequency responsive", and which could be more amenable to analysis in the time-domain because the information obtained relates directly to the performance of the system in time. In time-domain analysis, the properties of a linear system can be summarized by its impulse response function $v(t)$, as illustrated in Fig. 1(a). The reason for this is that an arbitrary input can be regarded as a linear combination of impulses occurring at various times, and by the assumption of linearity, the output is then that same linear combination of impulse response functions as shown in Fig. 1(b).

Such a situation may be described using the "systems approach", formally represented in Fig. 2. Here we have a linear, time-independent system with one measured input, and one measured output. Linear systems have the property that an input sequence $X(t)$, which produces an output sequence $Y(t)$, will produce an output sequence $aY(t)$ if the input sequence is $aX(t)$, where a is time independent. More generally, if a different input sequence $X'(t)$ produces an output sequence $Y'(t)$, then the combined input $(aX(t) + bX'(t))$ will produce an output sequence $(aY(t) + bY'(t))$. An assumption of a linear system may be unduly restrictive since many biological phenomena (e.g. bacterial population growth) are known to be non-linear. However, a non-linear system can appear linear for small changes in the inputs. From a mathematical standpoint, linear systems are much easier to analyse than non-linear systems, and it is therefore sensible to use a linear description for a first attempt.

The system in Fig. 2, with a single input and a single output, is just a particular example of the general multi-input multi-output system shown in Fig. 3. The measured inputs and outputs are labelled X_i and Y_i, respectively, the unmeasured inputs are labelled U_i and the measuring devices (or procedures) used to collect the data are labelled M_i. Underlying the scheme in Fig. 3 is the notion that changes in the inputs X_i will cause some changes in the outputs without affecting the other inputs. There will also be some changes in the outputs which are caused by changes in the unmeasured inputs and cannot, therefore, be traced to changes in the measured inputs. Such unexplainable variations in output are often referred to as "noise" and would be observed even if all the measurements were precise and error free. Of course measurement error will cause additional "noise".

A further possibility is that the properties of the system may be changing in time or in response to external conditions and stimuli. Such a possibility can be incorporated using the notion of a system "state". In Fig. 3 the system is imagined to be in a state α which is in effect a parameter relating the inputs and outputs. An example of this might be the differing responses of a physiological subject to a stimulus depending on the degree of anaesthesia. It is also possible that the system response may be time dependent, e.g. depending on the age of the physiological subject referred to in the previous example. Another useful application of the concept of system state is in analysing non-linear systems, for example, by making the system state dependent on the magnitude of the input. Such an approach may be likened to a piece-wise linear approximation to a non-linear function, a technique often used in electronic analogue circuits.

Most biological systems are multivariate and time dependent. For simplification purposes a traditional approach has been used to reduce the number of variables (e.g. using environmental chambers, water baths and other regulatory systems). Similarly the experiment can often be designed

7. THEORY AND PRACTICE OF TIME-DOMAIN TECHNIQUES

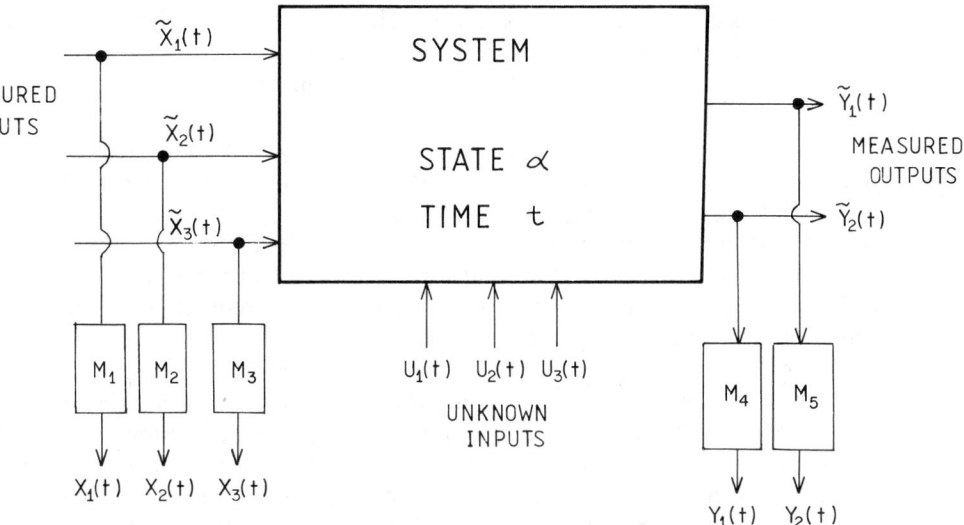

Fig. 3. A general linear system with several inputs and outputs.

to ensure that the system state does not change during the collection of data (e.g. using a chemostat culture). Although there is no inherent difficulty in treating multivariate systems, we shall restrict the following discussion to single input, single output systems.

2. Time-Domain Techniques

2.1. Introduction

The system modelling techniques of time-series analysis fall conveniently into two categories: first, those which produce models describing the structure of a single series, the system output; and second, those relating one series, the system output, to another series, the system input. The aim, for single series, is to construct a statistical model which incorporates the serial correlations, or relationships between successive points in the series. This procedure reduces the variance of the series, and one criterion for judging the efficiency of the chosen model is the variance reduction achieved. Similarly, the aim for models relating input and output is to fit an impulse-response model, and the efficacy of such a model can also be judged in terms of the variance reduction achieved from the output series. Clearly, if the variance of the residual series after fitting the model exceeds that of the original output series, the model is imposing structure on the output series. This situation indicates either that the structure of the model is wrong, or that there is no significant relationship between the output from the system and the supposed input.

Although the criterion of variance reduction is paramount, it is also important to restrict the number of parameters in the model. Clearly, the variance of the residual output series after model fitting can be made equal to zero by using one parameter for each data point fitted, but such profligate use of parameters is of no practical use. Thus a balance must be found between the conflicting demands of maximal variance reduction and parsimony in the number of parameters employed.

The impulse-response function, the key to the time-series modelling of relationships between input to and output from a system could in principle be obtained directly by imposing a unit impulse signal on the system input. However, the practical problems are usually sufficient to rule out this particular method. For example, it may not be possible to keep the input signal at zero or some constant value before and after the inpulse, and even if that were possible, the response at the output might be masked by variations unrelated to the input. Consequently, statistical techniques are required in the search for the impulse-response function of most practical systems.

2.2. The Box and Jenkins Techniques

The statistical techniques for the time-domain modelling of single series and relationships between series were formerly widely scattered in the literature, but have been brought together and extended by Box and Jenkins (1970). In so-called "single-series" analysis, a linear stochastic model can be applied to the time series consisting of the output of a system. Using the stochastic model in conjunction with the theory of forecasting presented by Box and Jenkins, it is possible to predict future values of the output series. This approach is attractive because it is necessary to collect time series at only one point in the system and because the analysis involved is relatively simple. However, the resulting single-series model depends entirely on the output data sequence used to fit the model. The model takes no account of the input to the system, and may fail to represent the output if the nature of the input changes.

A more satisfactory approach is to include the input series explicitly, and Box and Jenkins refer to this as "transfer-function" analysis. This term has also been used in frequency-domain and Laplace-transform analysis but no confusion need arise if one regards it as a general term covering all relationships between input and output for a linear system: the time-domain version considered here is just the impulse-response function referred to earlier. The transfer function, or impulse-response function, in principle gives a complete description of a linear system, and its nature is determined by the underlying processes within the system.

2.2.1. Modelling of Single Series

The analysis and description of the data set is carried out in two stages. The first stage is the identification stage, where the most appropriate type of model is selected, and the second stage is the estimation stage, where the parameters of the model are estimated and the goodness of fit of the model is tested.

First, let us make a few introductory remarks about time series. A time series of observations $X_t, X_{t+1}, \ldots$ is strictly stationary if all statistical properties of that series are invariant under shifts of observation time by an integral number of sampling time intervals. In particular, the mean, μ_x and variance σ_x^2 of a stationary series are constant, and may be estimated from any sequence of N values:

$$\hat{\mu}_x = \overline{X} = \frac{1}{N}\sum_{t=1}^{N} X_t \tag{1}$$

$$\hat{\sigma}_x^2 = S_x^2 = \frac{1}{N-1}\sum_{t=1}^{N}(X_t - \overline{X})^2 \tag{2}$$

Thus, a first check for stationarity can be carried out by splitting the data set into equally sized segments, or groups of sequential observations, and computing the mean and standard deviation of each segment. However, it is computationally simpler to consider the range R, rather than the standard deviation as a measure of "spread". Plotting R against the mean for all segments then gives an indication of the stationarity of the series. If the plotted points fail to cluster around the point $(\overline{X}, \overline{R})$ the series is said to be seriously non-stationary. Failure to cluster around the mean is normally obvious and is easily dealt with as will be shown later. Failure to cluster round the mean range, $\overline{R}$, implies that the series requires transformation. This is because non-stationarity in the variance may have the effect of "masking" recognizable patterns in the autocorrelation function and thereby hindering model identification. A logarithmic transformation or power-law transformation may be used to induce stationarity in variance, as illustrated in Fig. 4.

In subsequent analysis, much use is made of the autocovariance function

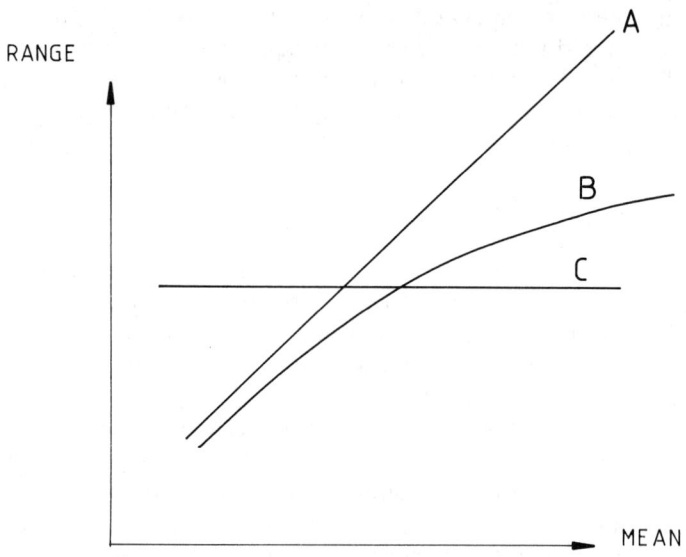

Fig. 4. Plots of range against mean and the appropriate data transformations. Case A: $T_x(X_t) = \log X_t$, Case B: $T_x(X_t) = X_t^g$ $(0 < g < 1)$, Case A: $T_x(X_t) = X_t$.

γ and the autocorrelation function ρ defined for some lag, τ, respectively by:

$$\gamma(\tau) = \text{cov}(X_t, X_{t+\tau}) = E\{(X_t - \mu_x)(X_{t+\tau} - \mu_x)\} \qquad (3)$$

$$\rho(\tau) = \frac{E\{(X_t - \mu_x)(X_{t+\tau} - \mu_x)\}}{\sqrt{(E\{(X_t - \mu_x)^2\}E\{(X_{t+\tau} - \mu_x)^2\})}} \qquad (4)$$

The property of stationarity gives

$$\rho(\tau) = \frac{\gamma(\tau)}{\gamma(0)} \qquad (5)$$

where

$$\gamma(0) = \sigma_x^2 \qquad (6)$$

A study of estimators $r(\tau)$ of the autocorrelation at delay τ is presented by Jenkins and Watts (1968). Here, the estimator used will be that favoured by Box and Jenkins (1970), namely:

$$r(\tau) = \frac{c(\tau)}{c(0)} \qquad (7)$$

In this eqn $c(\tau)$ denotes the estimator of the autocovariance, $\gamma(\tau)$:

$$c(\tau) = \frac{1}{N} \sum_{t=1}^{N-\tau} (X_{t+\tau} - \bar{X})(X_t - \bar{X}) \tag{8}$$

This estimator has the property that it tends to zero at large lags as the number of terms in the summation is reduced. For stationary data, the autocorrelation is a symmetric function normalized to unity for zero lag:

$$r(0) = 1$$

$$r(-\tau) = r(\tau)$$

The autocorrelation function thus defined is a fundamental tool in the analysis of time series and, in particular, in the determination of whether the series has been generated by an autoregressive or moving average process.

It is convenient to define a "backward step" operator, B, following the notation of Box and Jenkins, by

$$BX_t = X_{t-1}$$

so that

$$B^n X_t = X_{t-n}$$

and

$$(1 - B)X_t = X_t - X_{t-1}$$

The mixed autoregressive-moving-average model of a stationary series X_t is denoted by

$$\Phi(B)X_t = \Theta(B)a_t \tag{9a}$$

where a_t is a "white-noise" series devoid of serial correlation.

The above equation when expanded, gives:

$$X_t - \Phi_1 X_{t-1} - \Phi_2 X_{t-2} - \cdots$$
$$= a_t - \Theta_1 a_{t-1} - \Theta_2 a_{t-2} - \Theta_3 a_{t-3} - \cdots \tag{9b}$$

In the case where $\Theta_i = 0$ for all indices i, the model is purely autoregressive. Thus, if only the first p parameters Φ_i are non-zero, the process is an autoregressive process of order p (denoted AR(p)) and is given by:

$$\Phi(B)X_t = a_t \tag{10a}$$

or

$$X_t - \Phi_1 X_{t-1} - \Phi_2 X_{t-2} - \cdots - \Phi_p X_{t-p} = a_t \tag{10b}$$

Similarly, where $\Phi_i = 0$ for all values of i, the model is purely moving average. Thus, if only the first q parameters Θ_i are non-zero, the process is moving average of order q (denoted MA(q)) and is given by:

$$X_t = \Theta(B)a_t \tag{11a}$$

or

$$X_t = a_t - \Theta_1 a_{t-1} - \Theta_2 a_{t-2} - \cdots - \Theta_q a_{t-q} \tag{11b}$$

Consequently, a mixed process (denoted ARMA(p, q)) is then specified by two numbers: p and q referring, respectively, to the number of autoregressive and moving-average parameters.

In the above expressions it is assumed that $\bar{X}$ is zero. If the mean is non-zero initially, it can be made zero by a transformation:

$$x_t = X_t - \bar{X} \tag{12}$$

The principal tools for identifying a process are the sample autocorrelation function and the sample partial autocorrelation function.

Let us consider first the MA(q) process (eqns (11a) and (11b)) and assume that the mean has been subtracted (eqn (12)). The covariance (eqn (13)) is then

$$\gamma(\tau) = E\{X_t X_{t+\tau}\}$$

and since a_t is presumed to be uncorrelated white noise we have also the relation

$$E\{a_t a_{t'}\} = \begin{cases} \sigma_a^2, & t = t' \\ 0, & t \neq t' \end{cases}$$

Using these results we may then obtain an expression for the autocorrelation function:

$$\rho(\tau) = \frac{\gamma(\tau)}{\gamma(0)} = \begin{cases} \dfrac{-\Theta_\tau + \Theta_1 \Theta_{\tau+1} + \cdots + \Theta_{q-\tau}\Theta_q}{1 + \Theta_1^2 + \Theta_2^2 + \cdots + \Theta_q^2}, & (\tau \leq q) \\ 0, & (\tau > q) \end{cases} \tag{13}$$

from which it follows that $\gamma(\tau)$ and $\rho(\tau)$ are both zero for $|\tau| > q$. Hence, the order of an MA(q) process may be identified from a cut-off in the sample autocorrelation function estimated from eqns (7) and (8).

If we next consider the AR(p) process we find that there is no sudden cut-off in the autocorrelation function. Assuming eqns (10a) and (10b) and that $\bar{X}$ has been subtracted (eqn (12)), we may derive an expression for $\rho(\tau)$ using eqn (4):

$$\rho(\tau) = \Phi_1 \rho(\tau - 1) + \Phi_2 \rho(\tau - 2) + \cdots + \Phi_p \rho(\tau - p) \tag{14}$$

Equation (14) shows that the autocorrelation function persists for delays longer than p, and therefore cannot be used to identify the order of process. However, on the basis of eqn (14) we may define the partial autocorrelation function which does have a cut-off at delay p, and serves to identify the process.

If we imagine fitting AR(k) processes with $k = 1, 2, 3, \ldots$, we shall generate a series of equations analogous to eqn (14):

$$\rho(\tau) = \Phi_{k,1}\rho(\tau - 1) + \Phi_{k,2}\rho(\tau - 2) + \cdots + \Phi_{k,k}\rho(\tau - k) \quad (15)$$

where $\Phi_{k,i}$ is the autoregressive parameter for delay i. Allowing τ to range from 0 to $k - 1$ we obtain k linear equations, the Yule–Walker equations, which may be solved for $\Phi_{k,i}$. If k, the assumed order of the process were greater than p, the actual order, we would find that $\Phi_{k,i}$ would be zero for $i > p$. In particular, the coefficients $\Phi_{k,k}$ would be non-zero for $k \leq p$ and zero for $k > p$, and therefore suitable for identification. The coefficients $\Phi_{k,k}$ ($k = 0, 1, 2, \ldots$) by definition form the partial autocorrelation function (with $\Phi_{0,0} = 1$), and may be estimated using the sample autocorrelations $r(\tau)$ and either a least-squares fit or a solution of the Yule–Walker equations. By contrast with the AR process, the MA process has a partial autocorrelation function which persists at long delays.

In some cases, it is necessary in the interests of parsimony to introduce both autoregressive and moving-average terms of order p and q, respectively. However, for the ARMA(p, q) process neither the autocorrelation function nor the partial autocorrelation function has a sharp cut-off. The first approximation to the value of the pair (p, q) is obtained using both the autocorrelation and partial autocorrelation functions, but the identification procedures depend on the values and relative values of p and q, and on the values of $\Phi_1, \ldots, \Phi_p, \Theta_1, \ldots, \Theta_q$. More details may be found in Box and Jenkins (1970).

In summary, it is the structure of the autocorrelation and partial autocorrelation functions which indicates the approximate value of the pair (p, q). Some examples of typical autocorrelation and partial autocorrelation functions for autoregressive processes are shown in Fig. 5, while those for moving-average processes are shown in Fig. 6.

The identification of ARMA(p, q) processes depends on the series concerned being stationary in mean $\bar{X}$. An obvious example where this assumption does not hold is where the data shows a trend, and the effect of a trend in the data is to mask any other effect which may be detectable in the autocorrelation or partial autocorrelation function. To deal with non-stationarity in mean, it is convenient to introduce the difference

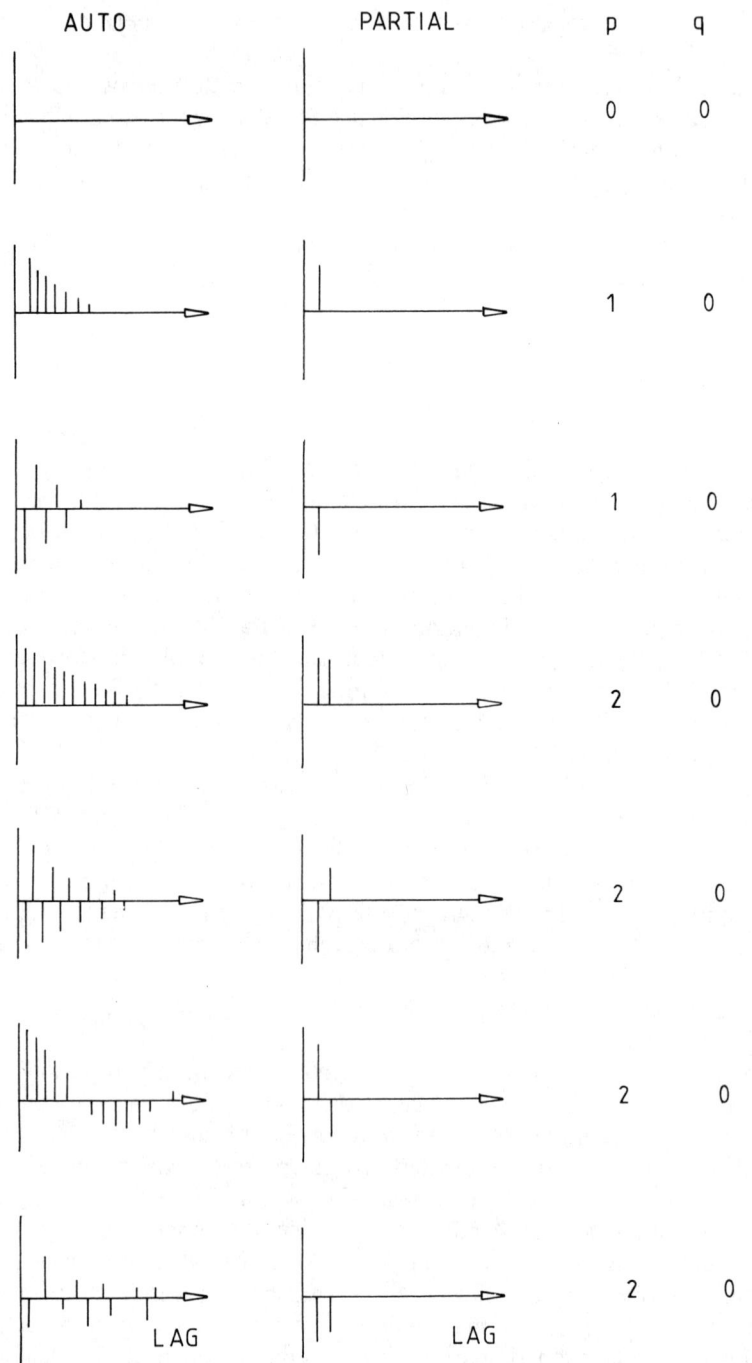

Fig. 5. Autocorrelation functions and partial autocorrelation functions for autoregressive processes.

7. THEORY AND PRACTICE OF TIME-DOMAIN TECHNIQUES

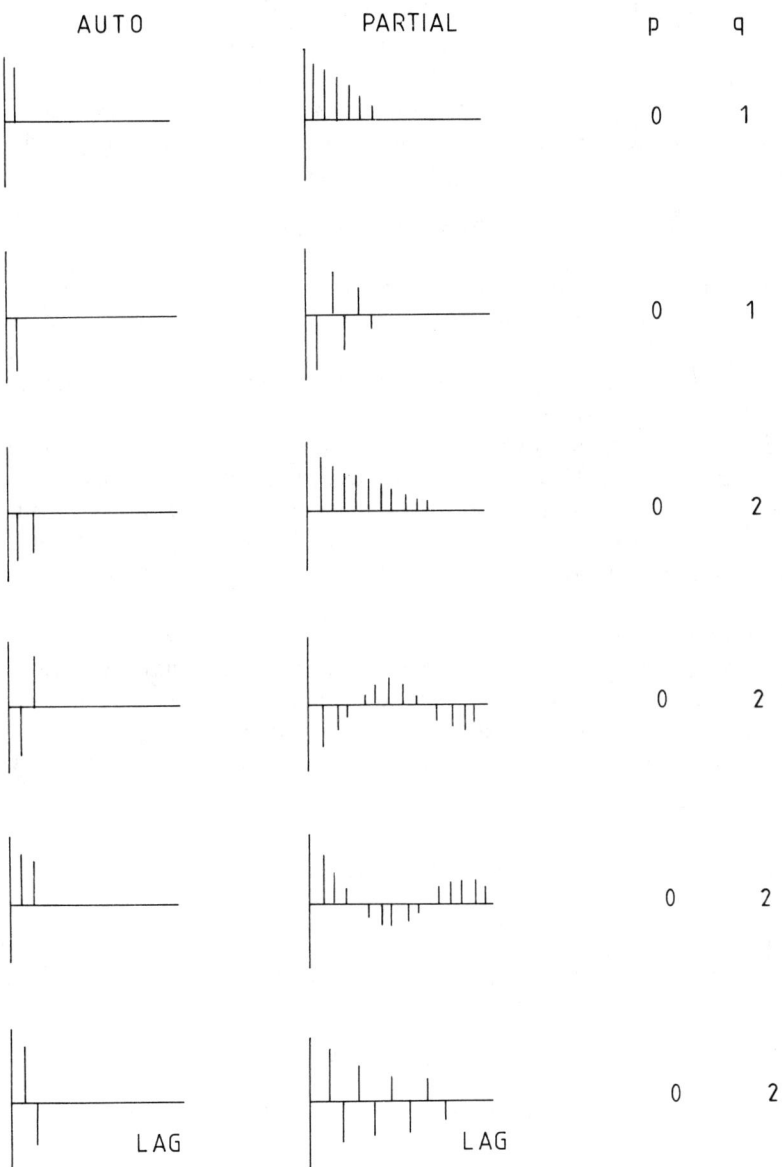

Fig. 6. Autocorrelation functions and partial autocorrelation functions for moving-average processes.

operator ∇ defined by:

$$\nabla = (1 - B) \tag{16}$$

The operator ∇^d may be used to induce stationarity in mean for the series. For example, a simple linear trend in a homogeneous data series can be removed using the operator ∇^d with $d = 1$, but a non-linear trend may require operation by ∇^d with $d = 2$ or more. Once the series has been rendered stationary, the procedure for a stationary series may yield the ARMA(p, q) model, resulting in the model

$$\Phi(B)\nabla^d x_t = \Theta(B)a_t \tag{17}$$

Since the non-stationary series is rendered stationary by using the dth difference, the series can be though of as having been generated by "integrating" the stationary process d times. The generating process is then described as an autoregressive integrated moving-average process of order (p, d, q) or ARIMA(p, d, q).

So far, it has been shown how a linear stochastic model can be identified for a given series. However, it was stated that non-stationarity can mask the characteristics of the model. The masking process can result from the presence of strong periodicities in the series, and consequently, it is necessary to obtain a model which describes the long-term periodic behaviour before attempting to model what is usually the smaller amplitude, short-term structure in the series.

Many phenomena exhibit seasonal or periodic behaviour, and such behaviour is often obvious when the data are plotted. Such periodic behaviour may be described by a seasonal model of period S denoted by ARIMA(P, D, Q)$_S$. The operator used in this circumstance is B^S giving

$$B^S X_t = X_{t-S}$$

The resulting seasonal model being ARIMA(P, D, Q)$_S$:

$$\phi(B^S)\nabla_S^D X_t = \theta(B^S)\alpha_t \tag{18}$$

and the residual noise series α_t is then described by a linear stochastic model ARIMA(p, d, q):

$$\Phi(B)\nabla^d \alpha_t = \Theta(B)a_t$$

This results in the general multiplicative model:

$$\Phi_p(B)\phi_P(B^S)\nabla^d \nabla_S^D X_t = \Theta_q(B)\theta_Q(B^S)a_t \tag{19}$$

The principles used in the identification and estimation stages are similar

7. THEORY AND PRACTICE OF TIME-DOMAIN TECHNIQUES 249

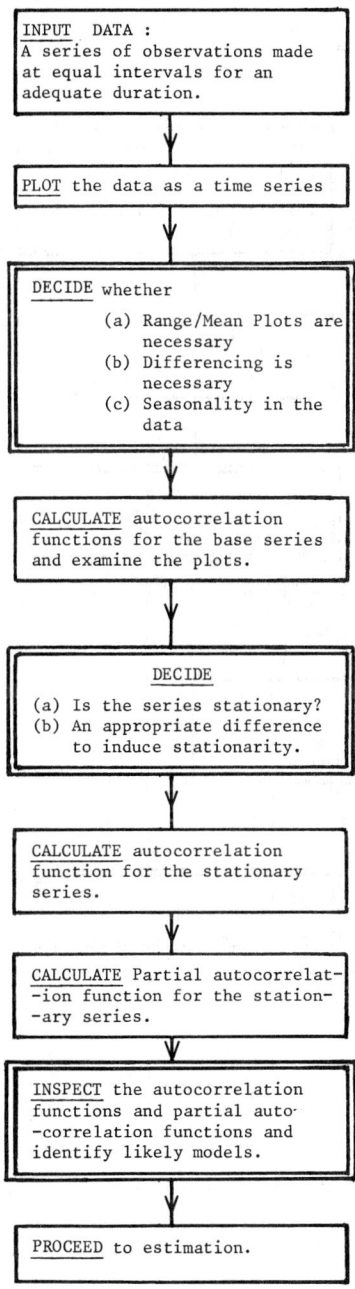

Fig. 7. The identification procedure for single-series models.

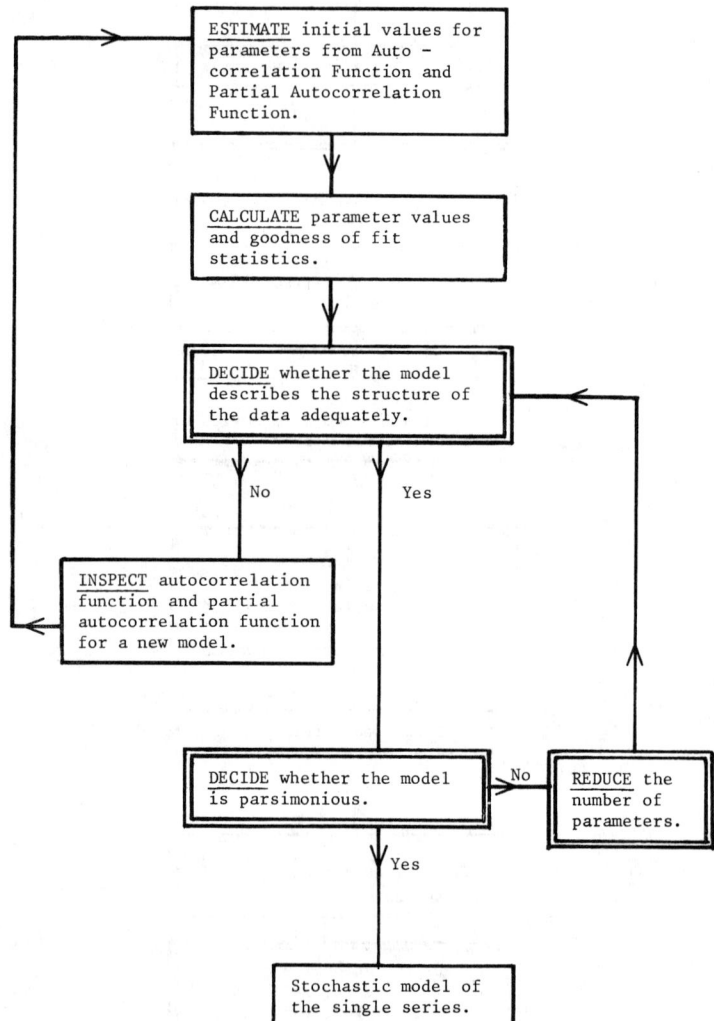

Fig. 8. The estimation procedure for single-series models.

to those outlined in the non-seasonal case, and are described by Box and Jenkins (1970).

The procedure for single-series model identification is summarized in Fig. 7, while the procedure for estimation of parameter values and final choice of model structure is summarized in Fig. 8.

2.2.2. Transfer-Function Models

For a linear system, the relationship between input series $\{X_t\}$ and output series $\{Y_t\}$ is governed by the impulse-response function, a linear filter of the form:

$$Y_t = v_0 X_t + v_1 X_{t-1} + v_2 X_{t-2} + \cdots$$

This relates the output at time t, Y_t, to the present and all previous inputs X_t by a convolution of the impulse-response function with the input series, X_t. In practice only a finite number of impulse-response function weights, v_i, are non-zero.

However, the above convolution depends on the relationship between X_t and Y_t being exact, which rarely occurs in practice. Consequently, some estimation of the possible error, or residual noise, must be included in the relationship. This results in the so-called transfer-function noise model of the system:

$$Y_t = v(B)X_t + N_t \qquad (20)$$

where

$$v(B) = (v_0 + v_1 B + v_2 B^2 + \cdots + v_L B^L)$$

and N_t may be regarded as a single series, described by an ARIMA(p, d, q) model as discussed in the previous section.

For the investigation of relationships between two series, the appropriate tools are the cross-covariance function for lag τ, denoted by $\gamma_{xy}(\tau)$ and given by:

$$\gamma_{xy}(\tau) = E\{(X_t - \mu_x)(Y_{t+\tau} - \mu_y)\} \quad (\tau = 0, 1, 2, \cdots) \qquad (21)$$

and the cross-correlation function at lag τ, denoted by $\rho_{xy}(\tau)$ and given by:

$$\rho_{xy}(\tau) = \frac{\gamma_{xy}(\tau)}{\sqrt{[\gamma_{xx}(0)\gamma_{yy}(0)]}} = \frac{\gamma_{xy}(\tau)}{\sigma_x \sigma_y} \qquad (22)$$

The estimator of $\rho_{xy}(\tau)$, the sample cross-correlation function $r_{xy}(\tau)$, is given by

$$r_{xy}(\tau) = \frac{C_{xy}(\tau)}{S_x S_y} \qquad (23)$$

The estimator of $\gamma_{xy}(\tau)$, the sample cross-covariance function, is given by

$$C_{xy}(\tau) = \frac{1}{N}\sum_{t=1}^{N}(X_t - \bar{X})(Y_{t+\tau} - \bar{Y}) \quad (\tau = 0, 1, 2, \cdots) \qquad (24)$$

once again following Box and Jenkins (1970).

Now, assuming that the input series $\{X_t\}$ can be described by a general ARIMA(p, d, q) model (disregarding seasonal components), let x_t be the stationary series

$$x_t = \nabla^d X_t$$

and y_t the series,

$$y_t = \nabla^d Y_t$$

Now

$$\Phi_x(B)x_t = \Theta_x(B)\alpha_t$$

so that (formally)

$$\Theta_x^{-1}(B)\Phi_x(B)x_t = \alpha_t \qquad (25)$$

where $\{\alpha_t\}$ is an uncorrelated white-noise series. By passing the output series through the same prewhitening filter, we obtain the series $\{\beta_t\}$

$$\beta_t = \Theta_x^{-1}(B)\Phi_x(B)y_t \qquad (26)$$

If the stationary input and output series are related by

$$y_t = v(B)x_t + n_t \qquad (27)$$

where n_t is the corrupting noise term, then

$$\beta_t = v(B)\alpha_t + \Theta_x^{-1}(B)\Phi_x(B)n_t$$

Multiplying throughout by $\alpha_{t-\tau}$ and evaluating the expectation, the values of the impulse-response function weights can be estimated from:

$$v_\tau = r_{\alpha\beta}(\tau)\frac{S_\beta}{S_\alpha} \qquad (28)$$

where S_α and S_β are the estimated standard deviations of the transformed input and output series, and $r_{\alpha\beta}(\tau)$ is the cross-correlation function for the series $\{\alpha_t\}$ and $\{\beta_t\}$.

7. THEORY AND PRACTICE OF TIME-DOMAIN TECHNIQUES

This is not a parsimonious representation, since it requires the estimation of all weights, $v_0, v_1, v_2, \ldots$, Box and Jenkins (1970) show, by induction, that a transfer-function model can be expressed parsimoniously in terms of a general linear difference equation:

$$(1 - \delta_1 B - \cdots - \delta_r B^r)Y_t = (\omega_0 - \omega_1 B - \cdots - \omega_s B^s)X_t \quad (29)$$

and for a system with "dead time" or "response time" equal to b sample time intervals, this becomes:

$$(1 - \delta_1 B - \cdots - \delta_r B^r)Y_t = (\omega_0 - \omega_1 B - \cdots - \omega_s B^s)X_{t-b} \quad (30)$$

i.e.

$$\delta(B)Y_t = \omega(B)X_{t-b}$$

where $\delta(B)$ and $\omega(B)$ are polynomials in B of order r and s, respectively. Substituting

$$Y_t = v(B)X_t$$

and ignoring for the moment the noise term, we obtain

$$\delta(B)v(B)X_t = \omega(B)B^b X_t \quad (31)$$

or

$$v(B) = \delta^{-1}(B)\omega(B)B^b$$

Equating coefficients of B in the equation

$$(1 - \delta_1 B - \cdots - \delta_r B^r)(v_0 + v_1 B + \cdots) = (\omega_0 - \omega_1 B - \cdots - \omega_s B^s)B^b$$

we see

$$\begin{aligned}
v_i &= 0 & (i < b) \\
v_i &= \delta_1 v_{i-1} + \cdots + \delta_r v_{i-r} + \omega_0 & (i = b) \\
v_i &= \delta_1 v_{i-1} + \cdots + \delta_r v_{i-r} - \omega_{i-b} & (b < i \leq b + s) \\
v_i &= \delta_1 v_{i-1} + \cdots + \delta_r v_{i-r} & (i > b + s)
\end{aligned} \quad (32)$$

Thus, by examination of the estimates of the impulse-response function weights, the estimate of the pure delay or "dead time" of the system is obtained. Using eqns (32) and Fig. 9, initial estimates of the values of r and s are obtained. Furthermore, preliminary estimates of the values of the parameters $\delta_1, \ldots, \delta_r, \omega_1, \ldots, \omega_s$ are obtained by solving the resulting simultaneous equations. The values of the parameters may then be optimized by least squares.

In spite of optimization, there will be an error term in the relationship

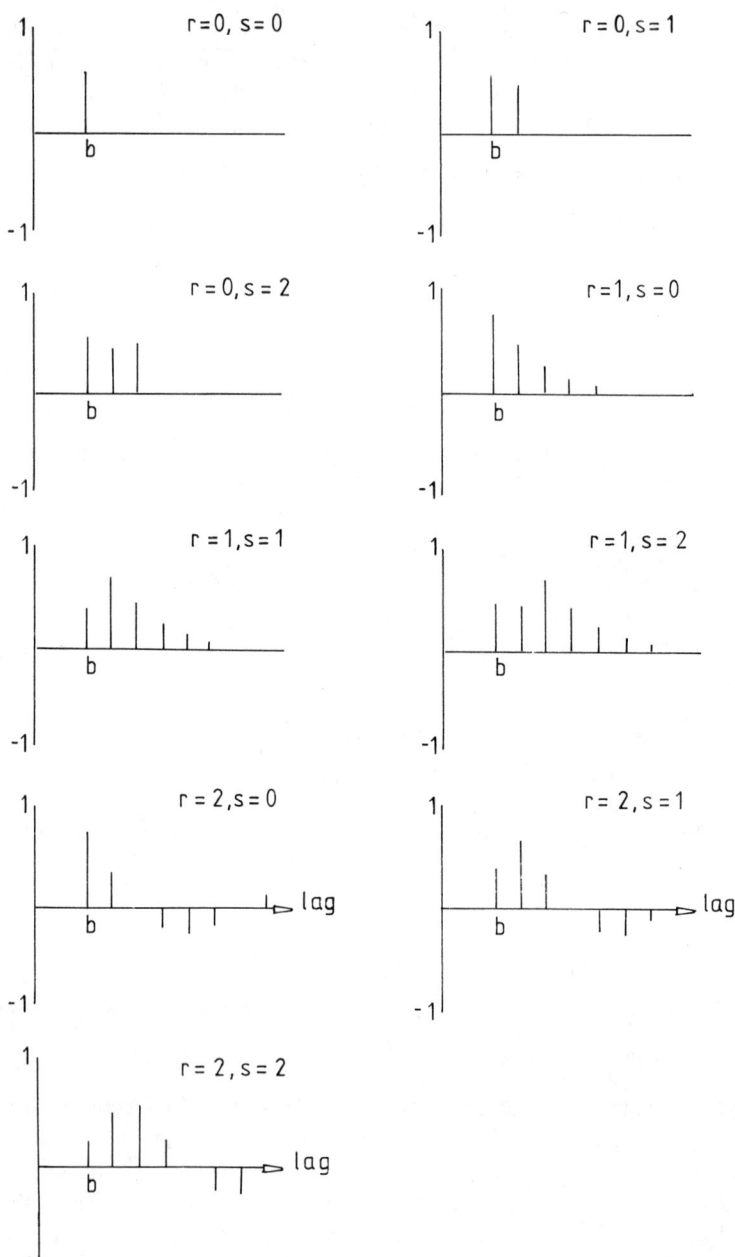

Fig. 9. Cross-correlation functions for different values of parameters *r* and *s* in a time-domain transfer-function model.

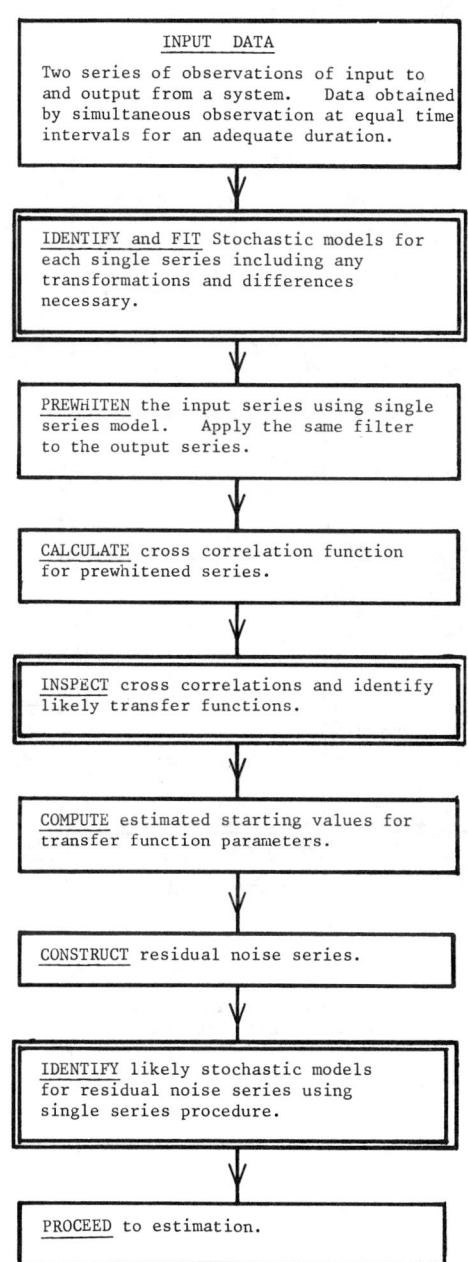

Fig. 10. The identification procedure for a transfer-function noise model.

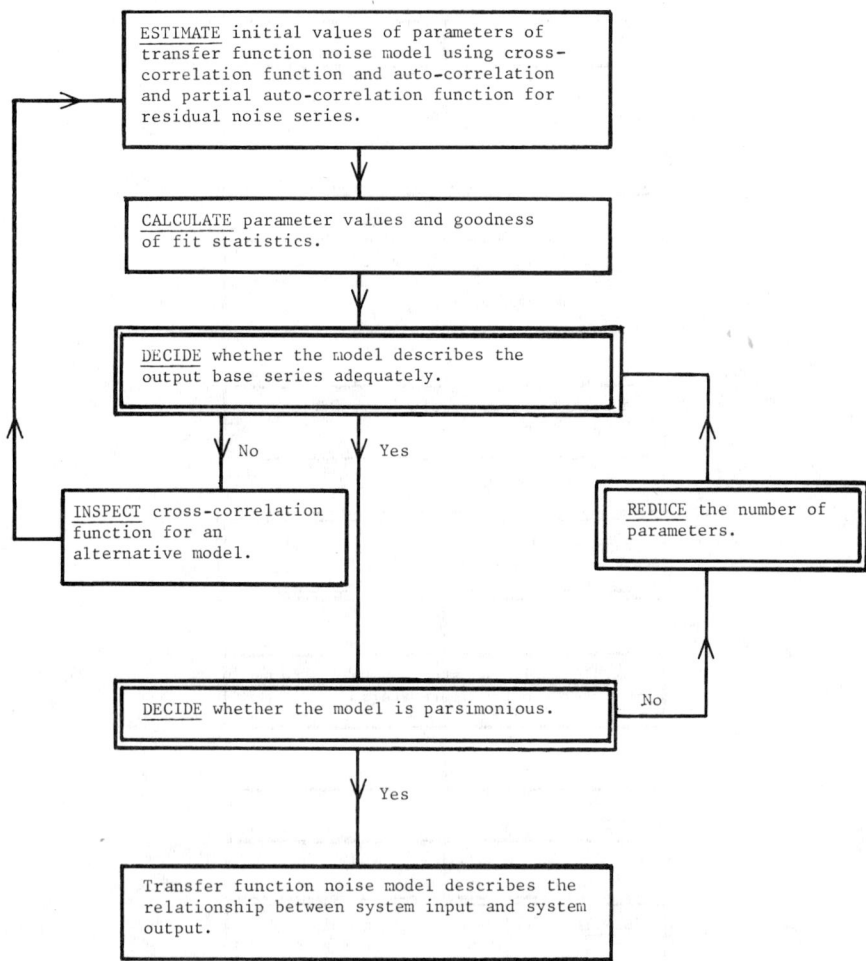

Fig. 11. The estimation procedure for transfer-function noise models.

given by

$$n_t = y_t - \delta^{-1}(B)\omega(B)B^b x_t$$

In order to complete the transfer function noise model, it is necessary to obtain a single-series description of the noise term, n_t. It may be assumed that there will be no seasonal component in the noise term, n_t, since this would form the gross structure of the transfer-function model. Thus, the noise term, n_t, can be represented by:

$$\Phi(B)n_t = \theta(B)a_t \tag{33}$$

The non-seasonal difference term ∇^d is normally incorporated in the transfer-function model, since non-stationarity in mean in the noise term implies non-stationarity in mean in the input or output series.

Thus, for two general series X'_t and Y'_t the most general form of transfer function between X'_t and Y'_t may be expressed parsimoniously as follows:

$$\nabla^d \nabla_S^{ds} T_y(Y'_t) = \frac{\omega(B)}{\delta(B)} \nabla^d \nabla_S^{ds} T_x(X'_{t-b}) + \frac{\Theta(B)}{\Phi(B)} a_t \tag{34}$$

$T_x(X'_t)$ and $T_y(Y'_t)$ correspond to the transformation required for X'_t and Y'_t to induce stationarity in variance of X'_t and Y'_t (cf. Fig. 4).

The operators ∇^d and ∇_S^{ds} are the non-seasonal and seasonal differencing operators respectively, S denoting the period of the seasonal components. $\omega(B)/\delta(B)$ is the transfer function of the system for which X'_t is the input and Y'_t is the output. $\omega(B)$ and $\delta(B)$ are polynomials in B of order r and s, respectively. $\Theta(B)/\Phi(B)$ is the transfer function of the filter which, when the input is white noise, a_t, the output is

$$n_t = Y''_t - \delta^{-1}(B)\omega(B)X''_{t-b} \tag{35}$$

where

$$X''_t = \nabla^d \nabla_S^{ds} T_x(X'_t)$$

and

$$Y''_t = \nabla^d \nabla_S^{ds} T_y(Y'_t)$$

The procedure for the identification of a transfer-function noise model is summarized in Fig. 10. The iterative procedure for estimation of parameter values and final choice of the parsimonious transfer-function noise model is summarized in Fig. 11.

3. Application of Time-Domain Techniques

3.1. General Considerations

3.1.1. *Data Sets*

Number of observations. One of the most important requirements for success in using time-domain techniques is that a suitable data set should exist. We have stressed earlier the requirement for data collected at equal intervals of time. The number of observations required depends on the nature of the data. For single-series model identification and estimation, the minimum number of observations should be in the region of three times the seasonal period plus 40 observations. For daily data, with a seasonal component with a period of a week, this would make the minimum number of observations approximately 60. In the absence of a seasonal component, this requirement can be reduced to about 40 observations.

In the case of transfer-function noise modelling, the requirement for data sets exhibiting seasonal behaviour is the "dead time" or response time of the system plus three times the seasonal period plus 40 observations.

Consider a system with a response time of 4 days, a seasonal component with period 1 week, sampled at daily intervals. The identification and estimation of a transfer-function noise model for such a system would require a minimum of about 65 data points. For non-seasonal data this can be relaxed to about dead time plus 40 observations.

Although these requirements for minimum numbers of observations in data sets can be taken as a guide, they do not preclude the analysis of time series of shorter duration. Shorter series can be analysed, but it is more difficult to identify the form of model required and to ensure that the model, once fitted, describes the relationships within the data set in the case of single-series models and between the data sets in the case of transfer-function noise models. It may be that, for shorter series than the minima suggested, several different models may describe the relationships adequately, but there may be a high degree of correlation between the parameters of any individual model. Shorter series also result in greater uncertainty in the estimated parameter values, and, as the number of degrees of freedom of the χ^2 goodness of fit test decreases, the test becomes less sensitive.

These considerations must all be borne in mind before attempting to use these techniques on small data sets. In the case of transfer-function noise modelling, some further constraints must be satisfied. Since we are considering the relationship between two different series, these series must have the same starting point and the same number of observations.

Transformations of the data. There are many instances where there is clearly a linear relationship between two series in their measured form. In such cases, we can proceed with transfer-function noise modelling in the way described.

However, other cases do exist where, either as a result of prior knowledge, or on inspection of the data, one must conclude that one or other, or indeed both series must be transformed in order to make use of the linear relationship between the values of the transformed variable.

Before embarking on any kind of transformation of the base series, it is always prudent to consider the implications that any transformation will have for the resulting models and the subsequent interpretation of the parameters of the models and their values.

Unequal time intervals. There is considerable interest in the application of time-domain techniques to problems where observations cannot be made at equal intervals of time. Such data present no problems where regression analysis is being used, but the same cannot be said for time-domain techniques.

All of the methods employed to make such data amenable to time-series analysis require that a smoothed continuous approximation be made to the observations. This approximation is then sampled to construct a time series. Wold (1974) shows how splines can be used to provide a continuous approximation to the data.

The drawback in using smoothed approximations is that all such techniques introduce serial correlation between successive points. Since a fundamental part of time-series analysis lies in describing serial correlation in the time series, it would be pointless to invest time in modelling features which were peculiar to the method used to generate a smoothed approximation to the data.

Nevertheless, a subset of the data collected at unequal intervals of time can be treated. This is the subset which arises from lost data. Small blocks of up to two or three missing data can be accommodated by using linear interpolation methods. Hamilton (1976) investigated the effect on the autocorrelation function of a variety of interpolated values for missing data, and it seems that there will be little impact on the structure of the models provided that no more than 5% of the data is missing.

3.1.2. Applications of Time-Series Analysis

The study of time series has preoccupied scientists and statisticians since before the turn of the century. However, the techniques have only really become viable with the advent of digital computers. Much of the early work was concentrated on studies in the frequency domain, and many workers continue to adopt this approach. The use of frequency-domain techniques has proved particularly fruitful in the investigation and modelling of the response of muscle and other tissue to various stimuli. Basar et al. (1974), for example, used autocorrelation and power-spectrum techniques to investigate the spontaneous activity of guinea pig taenia coi. Lucas and Harper (1976) used frequency-domain techniques to investigate periodicities in demand for electrical stimulation to maintain wakefulness in cats, and Chess et al. (1975) used periodic brain stimuli followed by time-series analysis to construct a mathematical model of the sympathetic heart response of cats.

There are systems, however, to which stimuli cannot easily be applied. This may be because the system is too large, e.g. macrometeorological systems, or because the system is only amenable to observation, e.g. fish catches. Nevertheless, time-series analysis of such systems has proved useful. Jackson et al. (1973) used time-series techniques to model relationships between climatological variables and hydrological and hydrogeological variables, and more recently, Ikeda and Stevenson (1978) analysed land- and sea-temperature time series collected by satellite and estimated time lags for these thermal systems.

Chock et al. (1975) used a similar approach to postulate relationships between meteorological variables and air quality data. A more specific application was the use by Kuehl et al. (1976) of time-series techniques to investigate the relationship between the boll retention in cotton plants and the 5-day average minimum temperature during the growing season. Hacker et al. (1973) derived relationships between meteorological data and mosquito populations by investigating periodicities in each series. Alcaraz and Wagensberg (1978) used time-series techniques to detect cyclic behaviour of copepod populations and female proportion in the population. These workers found negative correlations at 30 days delay and positive correlations at 15 days delay.

Botsford and Wickham (1975) used correlation coefficients at various delays to obtain relationships between the upwelling index and Dungeness crab catches, and also investigated the cyclic nature of the crab catches at various sites.

Examples of the analysis of time series in the time domain are less plentiful, and the nature of stimuli is somewhat more complex in this case. Jensen (1976) analysed historical records of the Atlantic Menhaden catch and identified and fitted a second-order autoregressive model to the data. This model was used for forecasting, and confidence intervals for forecasts were found. Hunter and Noordergraaf (1976) compared the frequency-domain techniques commonly used in haemodynamics with time-domain techniques. These authors concluded that, when considering the heart, the time-domain impulse-response function seemed a more appropriate model than the electrical analogue impedance.

Some authors have used a variable other than time as the basis for their series. For example, Webster and Cuanalo (1975) used the methodology of time-series analysis to model the relationships between soil properties at different points along a linear transect.

Some further examples of the application of time-domain techniques to sewage treatment plant data can be found in Coackley *et al.* (1978), Berthouex *et al.* (1975) and Huck and Farquhar (1974).

3.2. A Worked Example

3.2.1. Introduction

The theory developed in the previous sections will now be applied to an example which shows the variability inherent in most biological systems.

The system which will be considered is a primary sedimentation tank of a municipal sewage-treatment plant as shown in Fig. 12. The sewage flow enters the cylindrical tank upwards at the centre, is forced downwards by a baffle, and leaves the tank by overflowing a circular exit weir. The designers intended that the sewage would take at least 2 h to flow through the tank and that a large proportion of the suspended particles would settle (in this relatively long time) to form a sludge which could then be disposed of by another process.

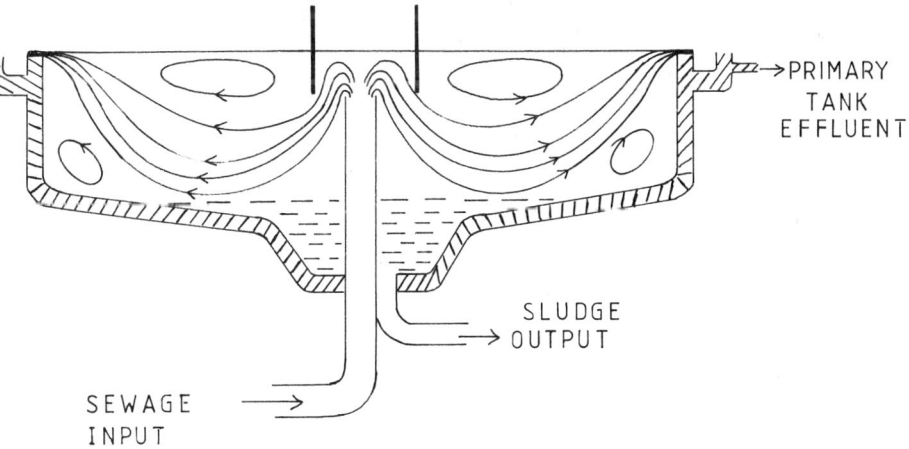

Fig. 12. A primary sedimentation tank of circular cross-section.

The efficiency of the sedimentation process is most directly studied by measuring the suspended solids concentration (S.S.) of the incoming sewage, X_t, and the primary tank effluent, Y_t. The sludge output can be ignored as a hydraulic factor since the average flow-rate will always be much less than the flow-rate of incoming sewage. Similarly the suspended solids concentration of the sludge may be ignored since its average value may in principle be determined from a mass balance calculated for the tank. The system can therefore be regarded as equivalent to Fig. 2 with a single input and single output. The state of the system may be regarded as determined by the hydraulic conditions, i.e. the incoming sewage flow-rate. For a single-state model to be appropriate the flow-rate should be approximately constant.

From engineering considerations it is reasonable to expect that the sewage will take between 0 and 12 h to pass through the tank, and so sampling at hourly intervals is indicated. The duration of observation would then have to be at least 3 days giving 72 samples of incoming sewage and

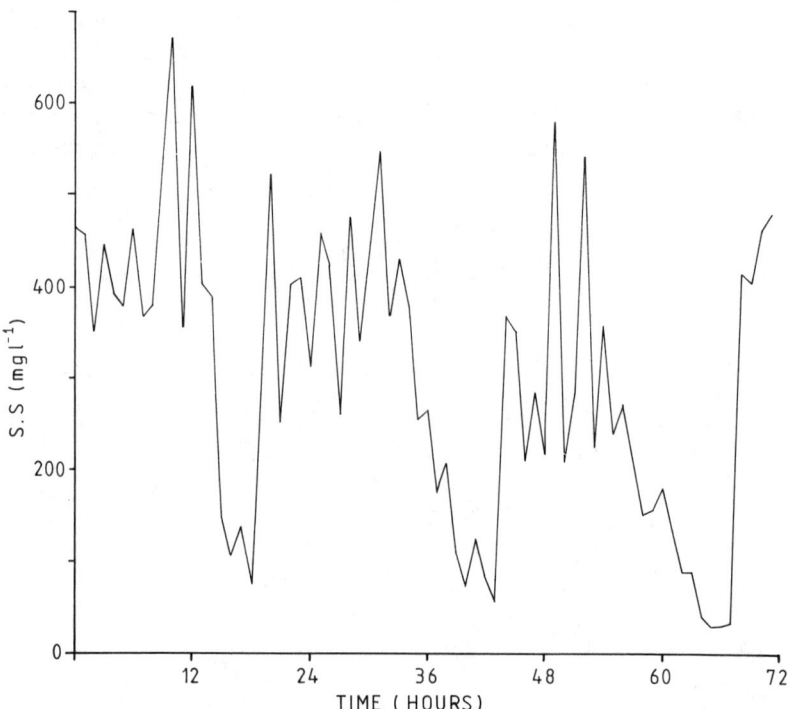

Fig. 13. The input series: suspended solids concentration of the incoming sewage.

7. THEORY AND PRACTICE OF TIME-DOMAIN TECHNIQUES

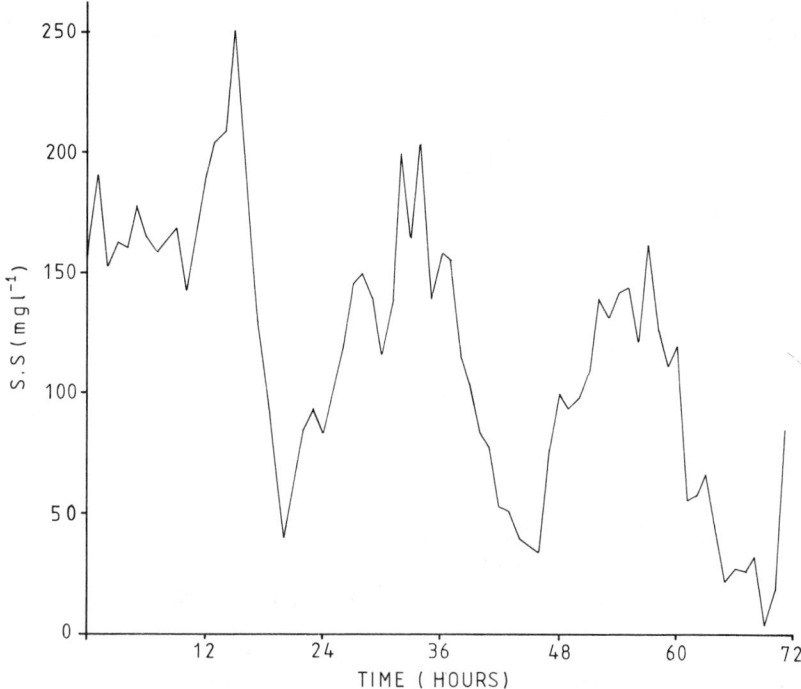

Fig. 14. The output series: suspended solids concentration of the primary tank effluent.

72 samples of primary tank effluent. Figures 13 and 14 show data collected in such an experiment.

3.2.2. Single-Series Models

Identification. The first stage in the identification procedure is to examine a plot of the data shown in Fig. 13 for the input series $\{X_t\}$ and in Fig. 14 for the output series $\{Y_t\}$. Both series show evidence of diurnal variation with superimposed fluctuations. Measurement error was about ±5% and so the fluctuations are not attributable to this cause. The mean values for the complete series are $\bar{X} = 304$ and $\bar{Y} = 116$, showing that the tank is removing about 62% of the suspended solids concentration. The standard deviations are $S_x = 159$ and $S_y = 55$, reflecting primarily the diurnal variation.

The two series are clearly not stationary in the means, and a "seasonal" model should be considered for both. The range-mean plots for 6 h segments

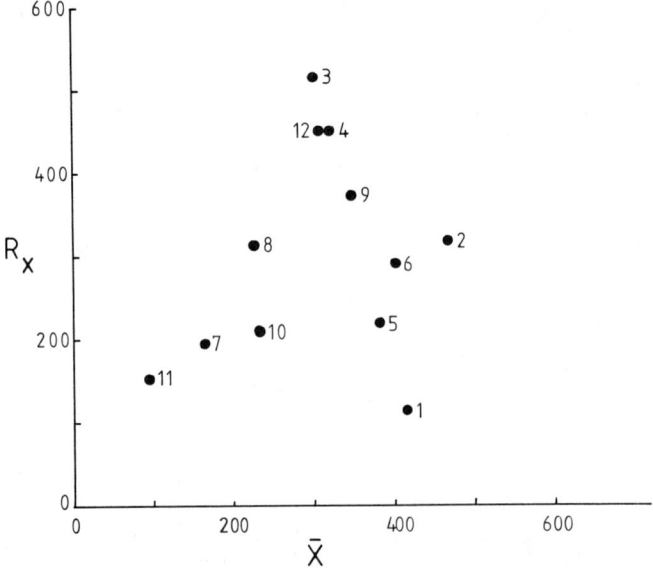

Fig. 15. Range-mean plot for the input series.

are shown in Figs 15 and 16 but in neither case is there a clustering around the centroid. However, no particular data transformation is indicated (cf. Fig. 4) and so we continue with the original series. The next stage is to examine the autocorrelation functions for the two series (Figs 17 and 18). In both cases a "seasonal" effect is clearly shown with a peak at a lag of

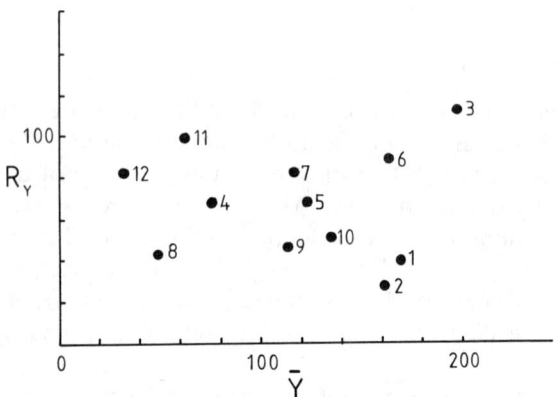

Fig. 16. Range-mean plot for the output series.

7. THEORY AND PRACTICE OF TIME-DOMAIN TECHNIQUES 265

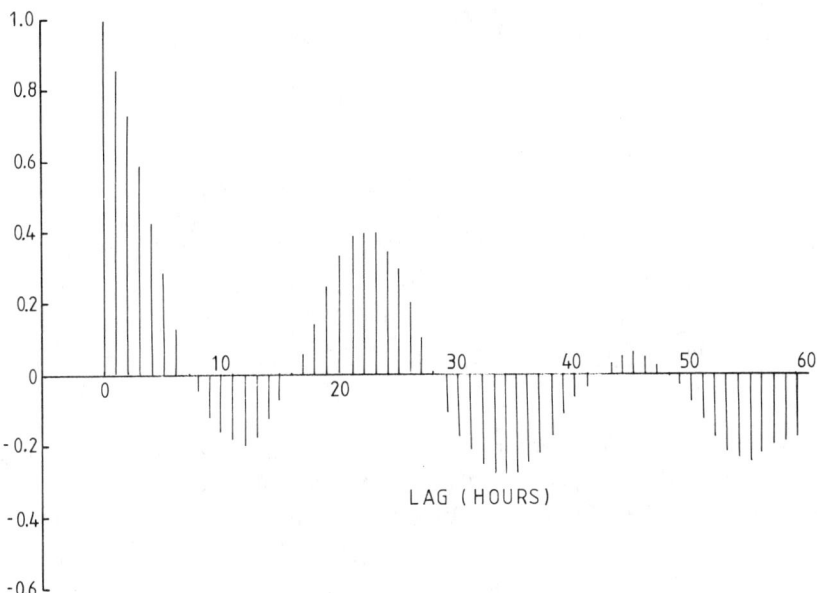

Fig. 17. Autocorrelation function of the input series.

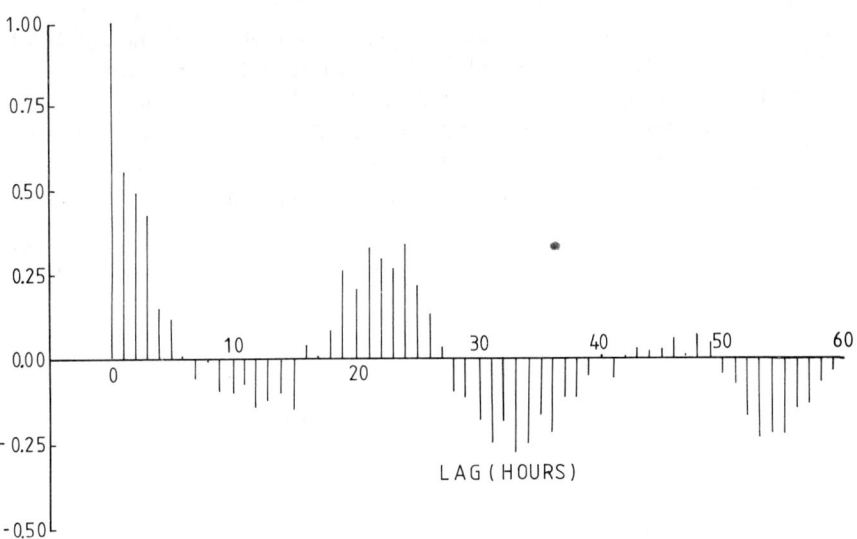

Fig. 18. Autocorrelation function of the output series.

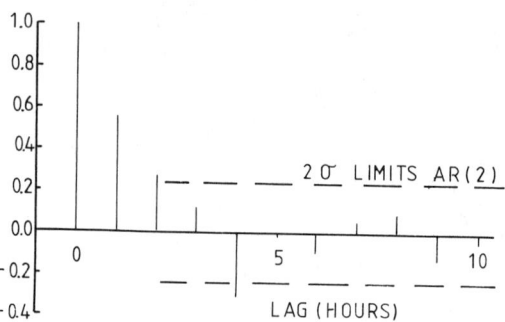

Fig. 19. Partial autocorrelation function of the input series.

between 22 and 23 h, rather than at 24 h as might have been expected. This curious result is probably due to the fact that we have only 3 days of data. A "seasonal" difference, using the operator

$$\nabla_S = (1 - B^S)$$

is obviously worth trying with $S = 22$, 23 and 24 h, in an attempt to induce stationarity in the two series. However, before we do that let us continue for the moment with non-seasonal models and examine the partial autocorrelation functions (Figs 19 and 20) for the undifferenced series.

The dashed lines in Figs 19 and 20 are based on a standard error $\hat{\sigma} = N^{-1/2}$ for partial autocorrelation coefficients at lags higher than the supposed cut-off. For the output series, AR(1) is indicated, but for the input series the situation is ambiguous with AR(2) or AR(4) as possibilities.

Let us consider the primary tank effluent (output) series, suspected to be an AR(1) process. The Yule–Walker estimate for the parameter is $\Phi_1 = 0.87 \pm 0.06$, and the variance reduction achieved is 77%.

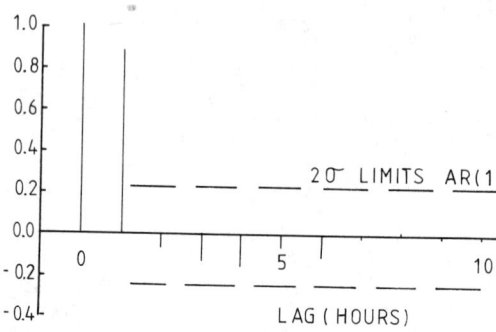

Fig. 20. Partial autocorrelation function of the output series.

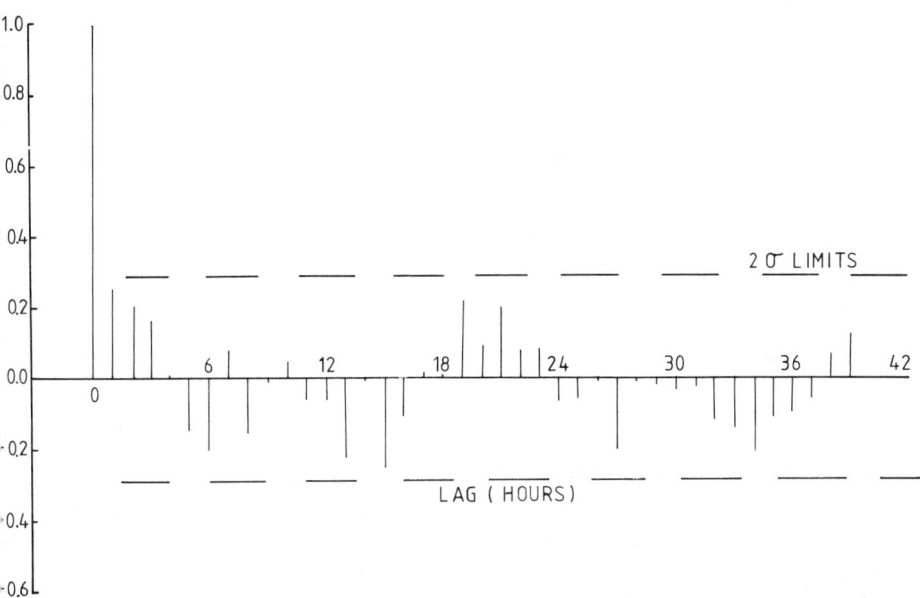

Fig. 21. Autocorrelation function of the input series after taking a 24-h difference.

Turning now to the incoming sewage (input) series, we find Yule–Walker estimates for an AR(2) process of $\Phi_1 = 0.40 \pm 0.11$, $\Phi_2 = 0.27 \pm 0.11$, and a correlation coefficient between the estimates of -0.55. The variance reduction with the above parameters is 37%. For the AR(4) model we find that Yule–Walker estimates give a variance reduction of 43%, but this is not significantly better than AR(2) as may be shown by using an F-test or the standard error of a difference method applied to the residual variances.

We consider next the possibility of making both series stationary in the mean by "seasonal" differencing. The most natural choice is a 24-h difference, which gives the autocorrelation function shown in Fig. 21 and a variance reduction (relative to the original series) of 34% for incoming sewage. Taking a 23- or 22-h difference gives significantly less variance reduction and, therefore, these possibilities need not be pursued further. Although Fig. 21 suggests that there is still some serial correlation after a 24-h difference the partial autocorrelation function (Fig. 22) shows no highly significant values. Turning to the possibility of a multiplicative model (eqn (19)), we face the difficulty of a data set too short for all but the first seasonal lag to be considered. However, it is found that a single seasonal autoregressive parameter gives more variance reduction than a seasonal

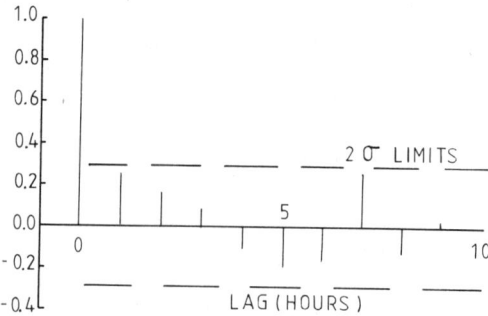

Fig. 22. Partial autocorrelation function of the input series after taking a 24-h difference.

difference, and when combined with two non-seasonal autoregressive parameters gives a variance reduction of 54%. Preliminary estimates of the parameters are: $\phi_S = 0.49 \pm 0.10$, $\Phi_1 = 0.29 \pm 0.14$, $\Phi_2 = 0.21 \pm 0.14$. The seasonal parameter is clearly significant and the variance reduction is significantly better than for the AR(2) model.

For the primary tank effluent series the variance reduction is 66% for a seasonal difference of 23 h, and 52% for a seasonal difference of 24 h. The partial autocorrelation function of the differenced series in both cases shows that an AR(1) model is required. However preliminary estimates show that the seasonal parameter ϕ_S is not significantly different from zero, and the total variance reduction achieved is 78% compared with 77% for an AR(1) model without a seasonal autoregressive parameter.

In summary, the following models have been identified: AR(1) for the primary tank effluent series, and a more complex model with one seasonal and two non-seasonal autoregressive parameters for the incoming sewage series. The models may be represented for incoming sewage S.S. as:

$$(1 - \Phi_1 B - \Phi_2 B^2)(1 - \phi_S B^{24})(X_t - \bar{X}) = a_t \qquad (36)$$

and for primary tank effluent S.S. as:

$$(1 - \Phi_1 B)(Y_t - \bar{Y}) = b_t \qquad (37)$$

where $\{a_t\}$ and $\{b_t\}$ are residual series.

Estimation and validation. The final stage in single-series modelling is to optimize the parameter estimates and then to check the residual series to make sure that the model is an adequate fit to the data. The optimization procedure will usually minimize the residual sum of squares function, which for the AR(1) model considered for the primary tank effluent series is

$$S(\Phi) = \sum_{t=2}^{N} (Y_t - \Phi Y_{t-1})^2 \qquad (38)$$

For autoregressive processes the minimization can be achieved using linear least-squares techniques (Bevington, 1969) which give the parameters as the (exact) solution of a set of linear equations. However, if there are moving-average parameters in the model the minimization becomes a non-linear least-squares problem (cf. eqn (39)) and must be achieved with an iterative procedure. A further difficulty with moving-average parameters arises because the initial values of the residuals are unknown. For example, in the MA(1) process $\{z_t\}$ we have:

$$a_t = z_t + \theta a_{t-1} = z_t + \theta z_{t-1} + \theta^2 a_{t-2}$$
$$= z_t + z_{t-1} + \theta^2 z_{t-2} + \theta^3 a_{t-3} \qquad (39)$$

which in principle must be iterated until it is deemed permissible to terminate the series by setting, say, $a_{t-m} = 0$. Such a procedure requires the first m values of z_t as starting values, where m might have a value of between 10 and 20. By contrast only one starting value is needed for the AR(1) model, as may be seen from eqn (38). Box and Jenkins (1970) discuss the use of "back forecasting" in which one starts with say the tenth point in the series and then estimates the earlier (unknown) values of a_t. This technique is particularly valuable with short series (say, less than 50 data points), but it must be recognized that the approximation relies on the values of parameters which are themselves being estimated. There is, therefore, a danger that the estimation procedure may not converge correctly, especially if the identification of the model is itself incorrect. However, in the present example back forecasting is unnecessary because the models are all autoregressive.

The results for the estimation of the models for the sedimentation tank input $\{X_t\}$ and output $\{Y_t\}$ are given in Table 1. The correlation matrices, Tables 2 and 3, do not show especially high correlation between the parameter estimates. If estimates were highly correlated, it would suggest that one or more of the parameters might be redundant, and that model simplification should be attempted.

As a further check on the fitted model, the residual series (Figs 23–25) and the residual autocorrelation functions (Figs 26–28) should now be examined. If the models have been identified and fitted correctly, the residual series should be uncorrelated "white noise". In particular the autocorrelation function should not be significantly different from zero, except of course for zero lag. For a finite series of n residuals, the standard

Table 1. Optimized single-series models.

Single-series model parameters	Incoming sewage S.S.		Primary tank effluent S.S.
	Seasonal	Non-seasonal	
ϕ_S	0.48 ± 0.14	—	—
Φ_1	0.36 ± 0.14	0.40 ± 0.11	0.875 ± 0.057
Φ_2	0.34 ± 0.14	0.29 ± 0.11	—
S_a	107.7	126.3	26.6
Q_{25}	20.1	36.6	21.0

Table 2. Correlation matrix of non-seasonal AR(2) model parameter estimates for incoming sewage S.S.

	Φ_1	Φ_2
Φ_1	1.00	−0.56
Φ_2		1.00

Table 3. Correlation matrix of seasonal model parameter estimates for incoming sewage S.S.

	ϕ_S	Φ_1	Φ_2
ϕ_S	1.00	−0.08	−0.11
Φ_1		1.00	−0.45
Φ_2			1.00

error of the autocorrelations will be approximately $\sigma = n^{-1/2}$ for all but the lowest lags. These 2σ-limits have been drawn on Figs 26–28 as dashed lines.

Box and Jenkins (1970) recommend what they call a "portmanteau" test of model adequacy based on the residual autocorrelation function $r(\tau)$. One calculates a quantity:

$$Q_K = n \sum_{\tau=1}^{K} [r(\tau)]^2 \qquad (40)$$

where K is typically about 25, and n is the number of time-series data used to fit the model. The value Q_K is then compared with the upper percentage

7. THEORY AND PRACTICE OF TIME-DOMAIN TECHNIQUES   271

Fig. 23. Residual series for the input-series AR(2) model

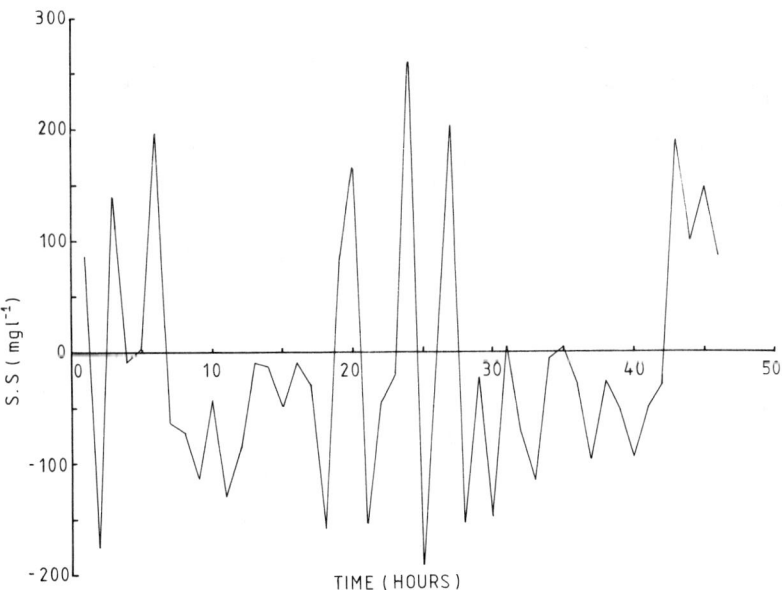

Fig. 24. Residual series for the input-series model with one seasonal and two non-seasonal autoregressive parameters.

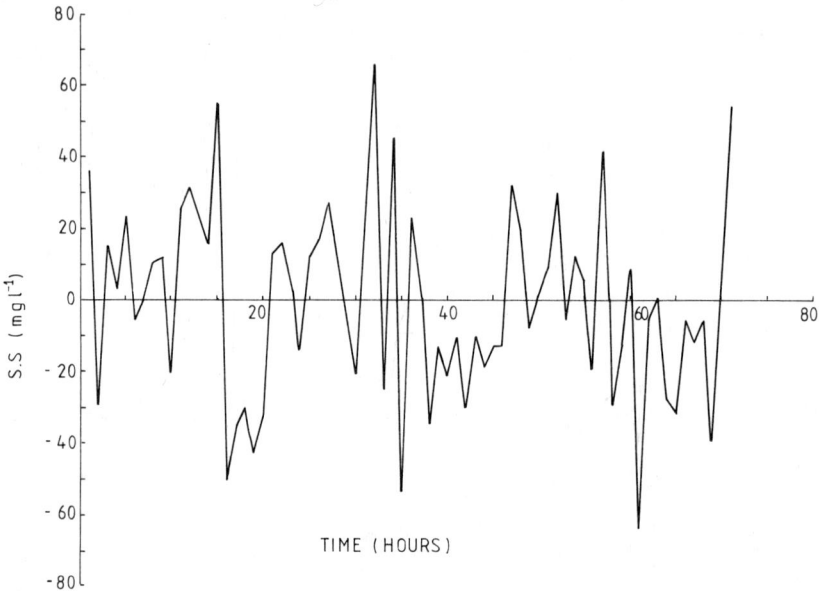

Fig. 25. Residual series for the output-series AR(1) model.

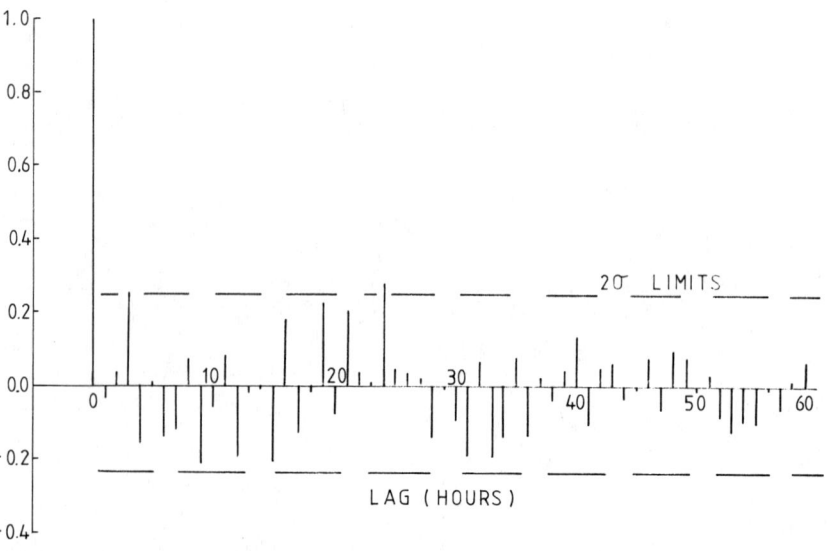

Fig. 26. Residual autocorrelation function for the input-series AR(2) model.

7. THEORY AND PRACTICE OF TIME-DOMAIN TECHNIQUES

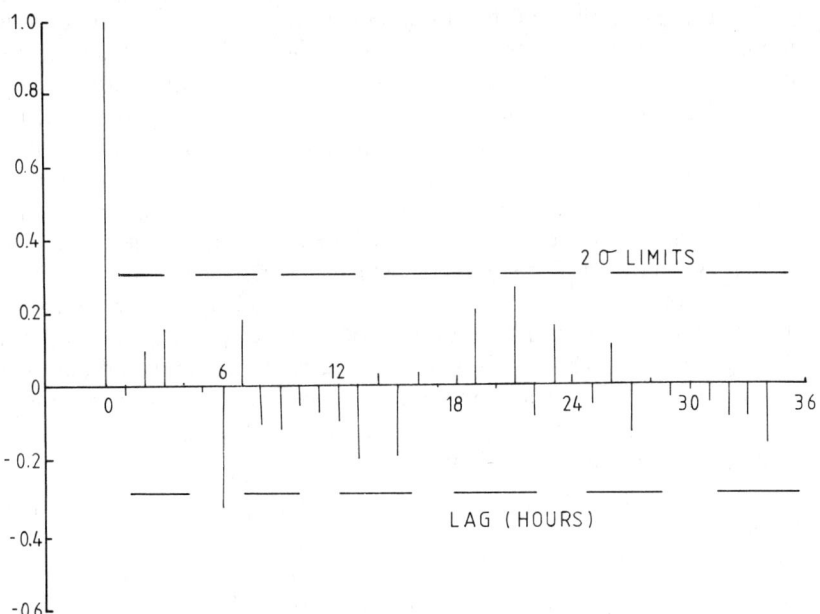

Fig. 27. Residual autocorrelation function for the input-series model with one seasonal and two non-seasonal autoregressive parameters.

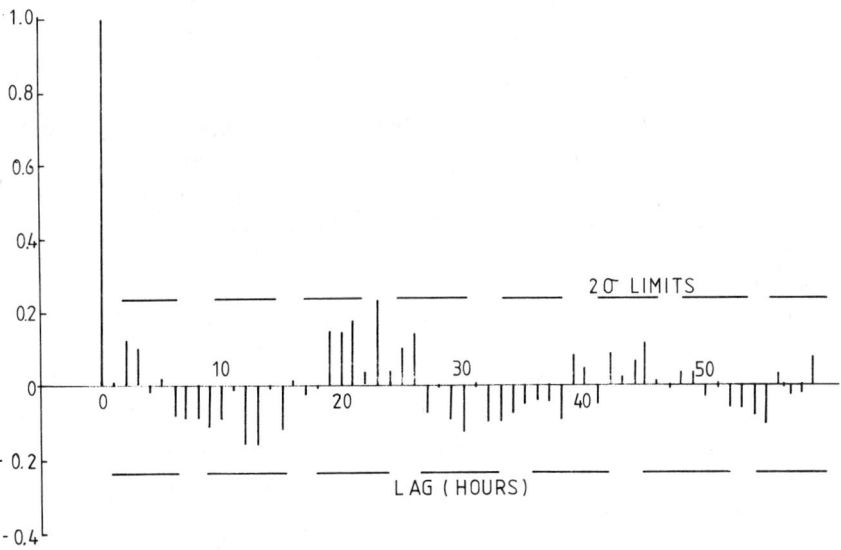

Fig. 28. Residual autocorrelation function for the output-series AR(1) model.

points of the χ^2 distribution for a number of degrees of freedom equal to K minus the number of parameters used in the model. If the model is a reasonably good fit one would expect Q_K to be approximately equal to the number of degrees of freedom. However, if the model is a poor fit, the value of Q_K will be much higher, and the tail area of the χ^2 distribution gives the probability of Q_K taking (by chance) that value or greater. Chatfield (1975) criticizes the test for its lack of sensitivity in discriminating between different models but it nevertheless is a useful guide. As an example, we show in Table 1 the seasonal and non-seasonal models for incoming sewage series. For the non-seasonal model $Q_{25} = 36.6$ compared with $\chi^2_{0.05,23} = 35.2$ whereas for the seasonal model $Q_{25} = 20.1$ compared with $\chi^2_{0.05,22} = 33.9$. This confirms that the seasonal model is the better of the two. For the primary tank effluent series we have $Q_{25} = 21.0$ compared with $\chi^2_{0.05,24} = 36.4$, indicating that the model is satisfactory.

The optimized models are then

$$(1 - 0.36B - 0.34B^2)(1 - 0.48B^{24})(X_t - \bar{X}) = a_t \tag{41}$$

for incoming sewage S.S. and

$$(1 - 0.87B)(Y_t - \bar{Y}) = b_t \tag{42}$$

for primary tank effluent S.S.

A less satisfactory model for the incoming sewage S.S. is

$$(1 - 0.40B - 0.29B^2)(X_t - \bar{X}) = a'_t \tag{43}$$

3.2.3. Transfer-Function Noise Model

Identification. The cross correlation between the input series $\{X_t\}$ and the output series $\{Y_t\}$ is shown in Fig. 29 and exhibits a peak at a lag of 4 h. Crowther *et al.* (1980) show that this peak should give a good approximation to the average retention time of the sewage in the sedimentation tank. However, for a more detailed description we try a prewhitening transformation.

For the seasonal model prewhitening transformation of eqn (41) applied to both the input and output series, the cross-correlation function $r_{\alpha\beta}(\tau)$ is given in Fig. 30. The only highly significant value occurs at a lag of 3 h. The cross correlations are consistently positive for lags up to 6 h, consistently negative from 11 to 18 h, and then positive from 21 to 29 h.

One disadvantage of using the model of eqn (41) for prewhitening is that the cross correlation is computed from only 46 points and has therefore a relatively large standard error. Using eqn (43) and the AR(2) model for prewhitening gives the cross correlation computed from 70 points and

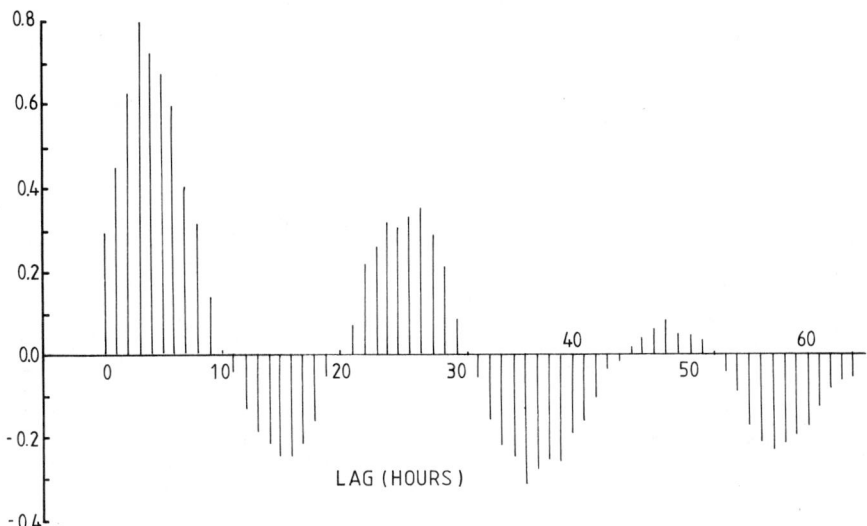

Fig. 29. Cross-correlation function between the input series and the output series.

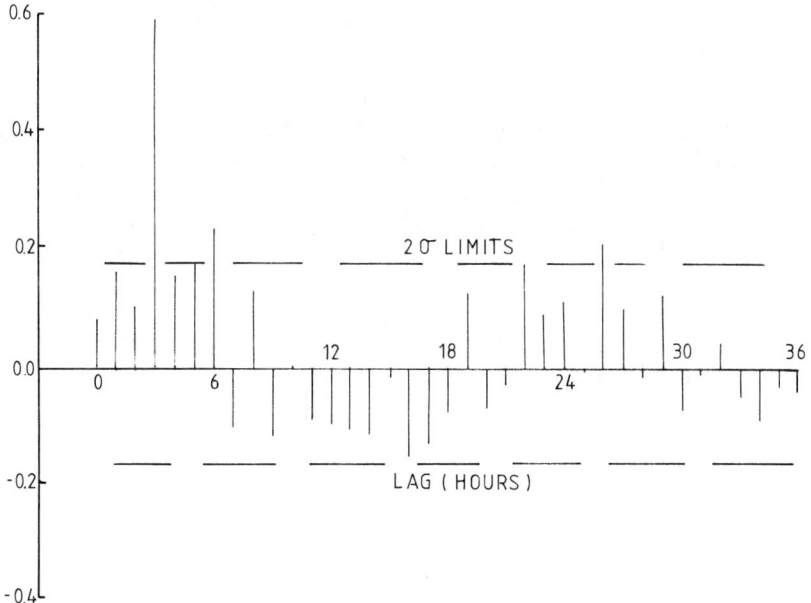

Fig. 30. Cross-correlation function between the input series and the output series after a prewhitening transformation with one seasonal and two non-seasonal autoregressive parameters.

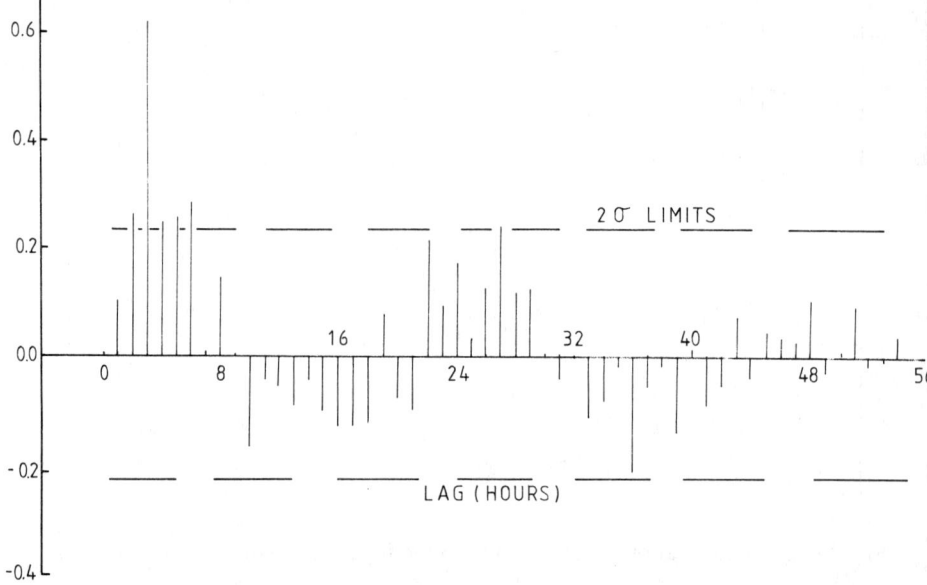

Fig. 31. Cross-correlation function between the input series and the output series after prewhitening using an AR(2) model.

shown in Fig. 31. Now the values are greater than the 2σ-limits for lags in the range 2–6 h. The only other value greater than these limits is at a lag of 27 h and this is undoubtedly linked by diurnal variation with the largest coefficient which is at the 3-h lag.

Thus neither prewhitening transformation gives a very clear identification

Table 4. Prewhitening estimates of the impulse-response coefficients.

Impulse-response coefficient V_1	Seasonal AR(2)	Non-seasonal AR(2)
V_0	0.021	0.000
V_1	0.045	0.027
V_2	0.027	0.063
V_3	0.156	0.147
V_4	0.042	0.058
V_5	0.047	0.061
V_6	0.063	0.067
V_7	−0.027	0.002
V_8	0.035	0.035

7. THEORY AND PRACTICE OF TIME-DOMAIN TECHNIQUES

of the impulse-response function. However, on physical grounds we would expect the impulse-response weights to be positive and non-zero for lags between 0 and 12 h at the most. A reasonable guess from Figs 28 and 29 would be to consider lags of 0–8 h. The estimates for the impulse-response function are then as given in Table 4, calculated from eqn (28).

From Table 4 it is not possible to identify r and s for a parsimonious transfer-function model (cf. Fig. 9). Under such circumstances it is in the authors' opinion worth trying to fit a non-parsimonious transfer-function model in which the impulse-response coefficients are parameters. In general this type of estimation is less efficient than using the parsimonious models of eqns (29) and (30), because there are more parameters. In Table 5 are shown the results of least-squares fits using different combinations of impulse-response weights and a single autoregressive parameter for the noise model. The models are thus of the form:

$$Y_t - \bar{Y} = \sum_i v_i(X_{t-i} - \bar{X}) + N_t$$

$$N_t = \Phi N_{t-1} + a_t \qquad (44)$$

where $\{a_t\}$ is the residual series. Only one autoregressive parameter is used because the single-series model already indicates this.

Table 5. Non-parsimonious least-squares estimates for the transfer-function noise model.

Impulse-response coefficients and autoregressive noise model parameter (eqn (44))	Model A	Model B	Model C
v_0	-0.032 ± 0.017	—	—
v_1	0.018 ± 0.016	—	—
v_2	0.078 ± 0.017	0.064 ± 0.016	0.066 ± 0.016
v_3	0.159 ± 0.018	0.155 ± 0.017	0.152 ± 0.018
v_4	0.072 ± 0.019	0.087 ± 0.018	0.086 ± 0.019
v_5	0.046 ± 0.019	0.052 ± 0.019	0.048 ± 0.020
v_6	0.034 ± 0.019	0.040 ± 0.020	0.047 ± 0.019
v_7	0.015 ± 0.019	-0.001 ± 0.018	0.005 ± 0.018
v_8	0.049 ± 0.018	0.033 ± 0.018	0.037 ± 0.018
v_9	0.013 ± 0.017	0.017 ± 0.018	—
v_{10}	-0.029 ± 0.017	—	—
Φ	0.235 ± 0.124	0.242 ± 0.124	—
S_a^2	228.2	252.1	

Table 6. Correlation matrix for model B in Table 5.

	V_2	V_3	V_4	V_5	V_6	V_7	V_8	Φ
V_2	1.00	−0.16	−0.30	−0.33	0.15	0.07	0.18	−0.07
V_3		1.00	0.06	0.26	0.33	0.18	0.05	0.00
V_4			1.00	0.00	−0.34	−0.37	0.11	0.03
V_5				1.00	0.02	−0.30	−0.39	0.02
V_6					1.00	−0.06	−0.31	−0.01
V_7						1.00	−0.06	−0.01
V_8							1.00	−0.01
Φ								1.00

The estimates in Table 5 all suggest the type of impulse response for $r = 1$, $s = 1$ and $b = 2$ in Fig. 9 and so we tentatively identify the transfer-function noise model using model B in Table 5. For completeness we show in Table 6 the correlation matrix for model B and in Fig. 32 the impulse-response estimates.

The preliminary estimates for the parsimonious transfer-function noise model are now obtained using eqns (32):

$$\omega_0 = 0.064, \quad \omega_1 = -0.166$$

$$\delta_1 = 0.566, \quad \Phi = 0.242$$

Estimation. The estimation stage for the transfer-function noise model is very similar to that for single-series models: first, the model parameters are optimized using a non-linear least-squares procedure, and second, the residual series is examined for evidence of lack of fit.

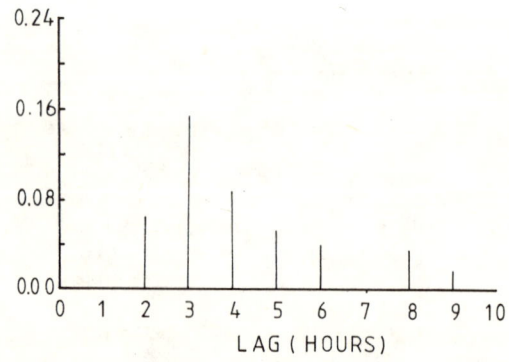

Fig. 32. Impulse-response estimates for model B, Table 6.

7. THEORY AND PRACTICE OF TIME-DOMAIN TECHNIQUES 279

Table 7. Optimized parsimonious transfer-function noise model.

Model parameter	Estimated value
δ_1	0.600 ± 0.042
ω_0	0.057 ± 0.016
ω_1	-0.121 ± 0.021
Φ	0.209 ± 0.125
S_a	16.47

Table 8. Correlation matrix for optimized parsimonious transfer-function noise model parameter estimates.

	δ_1	ω_0	ω_1	Φ
δ_1	1.00	0.07	0.68	-0.01
ω_0		1.00	0.67	-0.06
ω_1			1.00	-0.05
Φ				1.00

Table 9. Impulse-response for optimized transfer-function noise model.

Impulse-response coefficient v_I	
v_0	0.000 —
v_1	0.000 —
v_2	0.057 ± 0.016
v_3	0.155 ± 0.023
v_4	0.093 ± 0.015
v_5	0.056 ± 0.010
v_6	0.034 ± 0.006
v_7	0.020 ± 0.004
v_8	0.012 ± 0.002
v_9	0.007 ± 0.001
v_{10}	0.004 ± 0.001

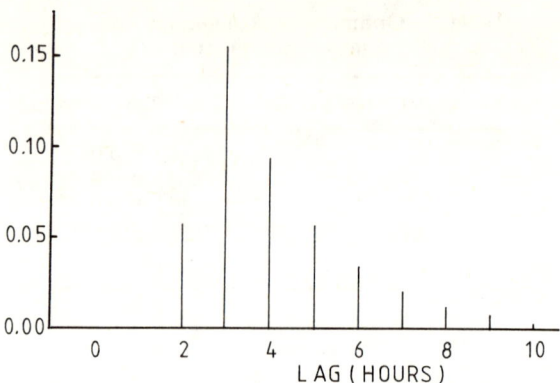

Fig. 33. Impulse response estimated using a parsimonious transfer-function model with $r = 1$, $s = 1$ and $b = 2$.

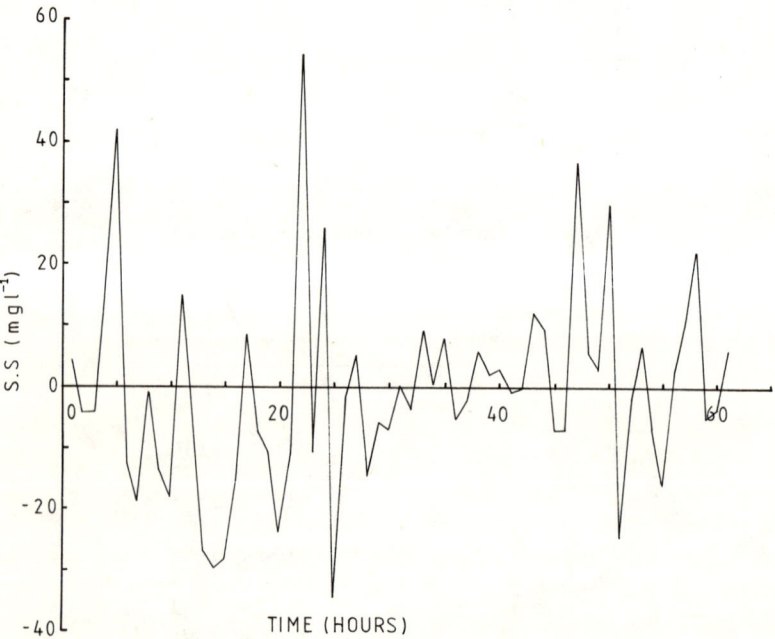

Fig. 34. Residual series for the parsimonious transfer-function noise model.

The optimized transfer-function noise model parameters are shown in Table 7 and their correlation matrix in Table 8. The model may be written as follows:

$$(Y_t - \bar{Y}) = (1 - 0.60B)^{-1}(0.06B^2 + 0.12B^3)$$
$$\times (X_t - \bar{X}) + (1 - 0.21B)^{-1}a_t \qquad (45)$$

and the impulse response is given in Table 9 and plotted in Fig. 33.
The residual series $\{a_t\}$ in eqn (45) is plotted in Fig. 34, and its auto-

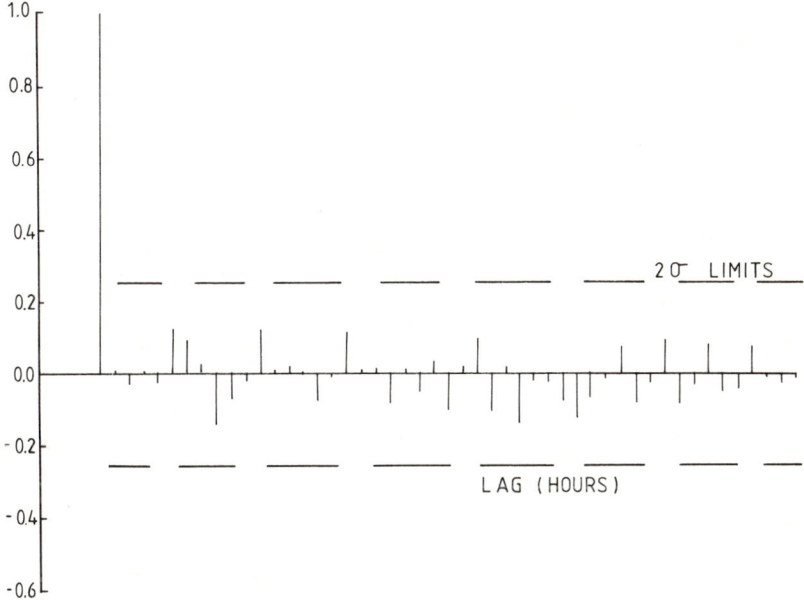

Fig. 35. Residual autocorrelation function for the parsimonious transfer-function noise model.

correlation function in Fig. 35. The portmanteau test of lack of fit gives $Q_{25} = 7.3$ compared with $\chi^2_{0.05,24} = 36.4$. For transfer-function models Box and Jenkins (1970) recommend checking the cross correlation (Fig. 36) of the residual series $\{a_t\}$ with the prewhitened input series $\{\alpha_t\}$. If this cross correlation is significantly different from zero it is likely that the model is deficient. Checking by cross correlation of the input series $\{X_t\}$ with the residuals $\{a_t\}$ is not recommended because, even if $\{a_t\}$ were uncorrelated white noise, the cross-correlation function would show a pattern similar to that of the autocorrelation of $\{X_t\}$. The cross correlation of residuals and prewhitened input (Fig. 36) shows that the model is acceptable. The portmanteau test of lack of fit may be applied to the cross correlations‡ and gives $Q_{25} = 13.7$ compared with $\chi^2_{0.5,23} = 35.2$ showing again that the model is acceptable. The correlation matrix in Table 8 shows a fairly high correlation between ω_1 and both ω_0 and δ_1, but this is to be expected given

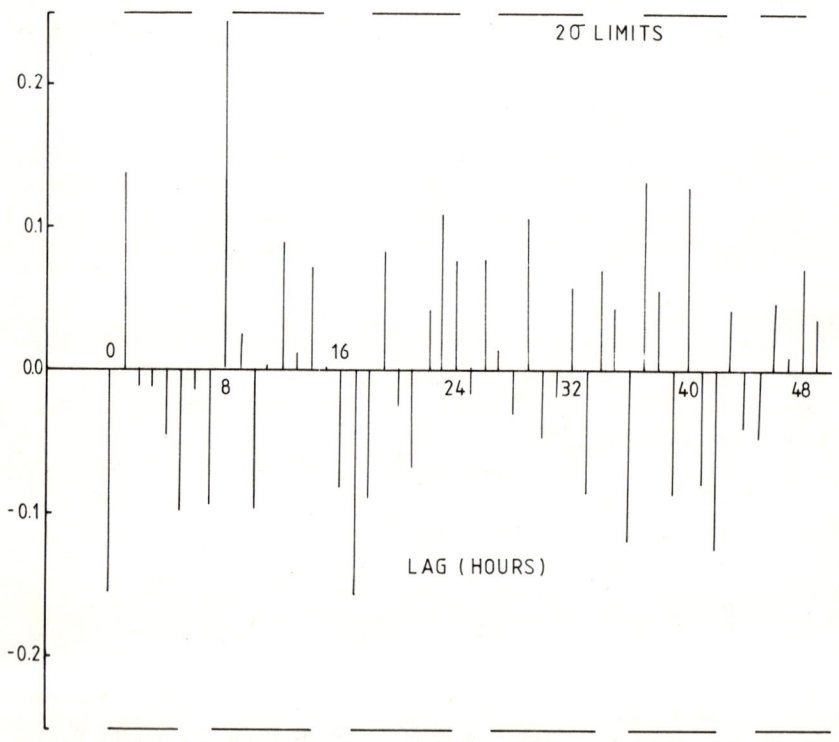

Fig. 36. Cross-correlation function between the AR(2) prewhitened input series and the residual series for the parsimonious transfer-function noise model.

the structure of the model (cf. eqns (32)) and given that there are now relatively few parameters. Hence we conclude that the transfer-function noise model is a satisfactory fit to the data, and we also note that the variance reduction relative to the output series $\{Y_t\}$ is 91% compared with 77% for an AR(1) single-series model.

‡ Starting at lag zero: i.e. Q_{25} includes lag zero to lag 24 h.

3.2.4. Discussion

The input series has a complex structure and it might be possible to find a better single-series model than those presented. The model with the seasonal autoregressive parameter fits the data better than the one without, but is less useful in obtaining preliminary estimates of the transfer-function model. With a longer data set it is likely that the seasonal model would perform better in that respect also.

By contrast, the modelling of the output series is straightforward and yields an AR(1) model. A tank with complete mixing would generate such an output process for a concentration variable (e.g. S.S.) if the input series were random and uncorrelated for that same concentration variable. This suggests that the impulse-response function has non-zero coefficients over a wide range of lags and tends to zero perhaps exponentially at high lags.

Fitting a transfer-function model using the impulse response coefficients as parameters shows a delay of 2 h ($b = 2$) and confirms that the impulse response tends fairly rapidly to zero by about a lag of 9 or 10 h (cf. Table 5 and Fig. 32).

A parsimonious model with just four parameters then gives an efficient fit to the data and results in the model shown in Table 9 and Fig. 33. From the impulse-response function one may obtain information about the system, e.g. the average retention time:

$$\bar{t} = \frac{\sum_{i=0}^{\infty} i v_i}{\sum_{i=0}^{\infty} v_i} \simeq 4.1 \text{ h}$$

This value agrees well with the lag of 4 h noted earlier for the first peak of the cross correlation between the input and output series (Fig. 29) as noted by Crowther et al. (1980).

The removal efficiency for suspended solids concentration is given by:

$$R = 1 - \sum_{i=0}^{\infty} v_i = 0.56$$

which is close to the corresponding figure obtainable from the mean values of incoming sewage S.S. and primary tank effluent S.S.

From the average hydraulic flow rate and the volume of the tank it is then possible to determine the "dead volume" of the tank, approximately 45% of the actual volume in this case (cf. Crowther et al., 1980).

Thus the fitted impulse-response function gives information on the dynamic performance of the tank under its normal operating conditions.

4. Conclusions

The Box and Jenkins techniques provide a wide range of models for describing both the serial correlations within a single time series and the correlations between two time series. Because the range of models is so wide, it is important to identify the most appropriate model and this is something of an art.

However, the artist's trick of observing only the most significant features of a scene by viewing through partially closed eyes is valuable in model identification also. In statistics this is known as significance testing. Having identified the model it is then necessary to check that it really does fit the data, and this again requires quantitative statistical tests.

The models once identified and fitted are essentially "black-box" models and require a biological or physical interpretation. However, the models, particularly for the impulse response, are a consequence of the biological and physical characteristics of the system. Hence, knowledge of the system can aid identification, and conversely, knowledge of the impulse-response function can assist in a study of the biology and physics of the system.

The three prerequisites for using Box and Jenkins techniques are:

(1) suitable data,
(2) the necessary computer programs,
(3) an understanding of the techniques.

Users are advised to use a well-known and well-documented package of programs (e.g. the widely available ISCOL package) because these will have been extensively tested. Box and Jenkins (1970) give data and corresponding analyses for users to test their programs, and this is especially recommended for those wishing to write their own. There are some disadvantages in using program packages because these often include more facilities than one needs and yet omit others that one would like, particularly high-resolution graphics. Perhaps, the best approach is a judicious blend of program from a package, supplemented by graphics and specialized programs written by the user.

References

Alcaraz, M. and Wagensberg, M. (1978). *Invest. pes.* **42**, 155–166.
Basar, E., Exroglu, C. and Ungan, P. (1974). *Pfluegers Arch. ges. Physiol.* **347**, 19–25.
Bendat, J. S. and Piersol, A. G. (1966). "Measurement and Analysis of Random Data." Wiley, New York.

Berthouex, P. M., Hunter, W. G., Pallesen, L. C. and Shih, C. Y. (1975). *J. Environ. Eng. Div., A.S.C.E.* **101**, 127–138.
Bevington, P. R. (1969). "Data Reduction and Error Analysis for the Physical Sciences." McGraw-Hill, New York.
Botsford, L. W. and Wickham, D. E. (1975). *U.S. Nat. Mar. Fish. Serv. Fish. Bull.* **73**, 901–907.
Box, G. E. P. and Jenkins, G. M. (1970). "Time Series Analysis Forecasting and Control." Holden-Day, San Francisco. (Reprinted with minor revisions (1976).)
Chatfield, C. (1975). "The Analysis of Time Series: Theory and Practice," p. 79. Chapman and Hall, London.
Chess, G. F., Varey, D. W., Henry, J. L. and Calaresu, F. R. (1975). *Ann. Biomed. Eng.* **3**, 189–198.
Chock, D. P., Terrell, T. R. and Levitt, S. B. (1975). *Atmos. Environ.* **9**, 978–989.
Coackley, P., Crowther, J. M., Hamilton, I. M. and Dalrymple, J. F. (1978). In "New Processes of Wastewater Treatment and Recovery" (G. Mattock, ed.), Chap. 4, pp. 61–74. Ellis Horwood, Chichester.
Crowther, J. M., Dalrymple, J. F., Woodhead, T., Coackley, P. and Hamilton, I. M. (1980). *Water Res.* **14**, 567–574.
Hacker, C. S., Scott, D. W. and Thompson, J. R. (1973). *J. Med. Entomol.* **10**, 533–543.
Hamilton, I. M. (1976). Ph.D. Thesis, University of Strathclyde, Glasgow.
Huck, P. M. and Farquhar, G. J. (1974). *J. Environ. Eng. Div., A.S.C.E.* **100**, 733–752.
Hunter, W. and Noordergraaf, A. (1976). *J. Appl. Physiol.* **40**, 250–252.
Ikeda, Y. and Stevenson, M. (1978). *Remote Sensing Environ.* **7**, 349–362.
Jackson, R. E., Gilliland, J. A. and Adamowski, K. (1973). *Water Resour. Res.* **9**, 1411–1419.
Jenkins, G. M. and Watts, D. G. (1968). "Spectral Analysis and its Applications." Holden-Day, San Francisco.
Jensen, A. L. (1976). *Chesapeake Sci.* **17**, 305–307.
Kuehl, R. O., Buxton, D. R. and Briggs, R. E. (1976). *Agron. J.* **68**, 491–495.
Lucas, E. A. and Harper, R. M. (1976). *Expl. Neurol.* **51**, 444–456.
Webster, R. and Cuanalo, De La C. H. E. (1975). *J. Soil Sci.* **26**, 176–194.
Wold, S. (1974). *Technometrics* **16**, 1–11.

Chapter 8

The Random Somatic Mutation Theory of Antibody Diversity and Information Theory

Alan Ebringer

*Immunology Unit, Departments of Biochemistry and Microbiology,
Queen Elizabeth College, London UK*

This final chapter differs from the rest of the book not just in its less than tenuous connection with microbiology but in the way that a mathematical theory has been applied to a biological problem. Alan Ebringer uses information theory, a branch of mathematics developed initially to investigate radio telephone signal transmission, to categorically refute an established theory of antibody diversity. In doing so he underlines the increasing necessity of exposing current biological dogma to the rigour that mathematical analyses are capable of imposing—*Editorial note.*

1. Introduction

A comprehensive theory of immunity must provide an answer to the two fundamental problems of immunology, that of antibody diversity and self-tolerance. First, it must provide a mechanism for generation of the large number of different antibody molecules which can be evoked against the enormous number of antigens found in nature and, second, it must provide a mechanism for the elimination or inactivation of antibodies which could potentially be made against self-antigens. The problem of self-tolerance was already perceived by Ehrlich at the turn of the century (Ehrlich and Morgenroth, 1901), when he coined the memorable phrase "Horror Autotoxicus" for the undesirable state which would supervene if antibodies were to be produced against self-antigens. The very existence of an antibody, a molecule specifically reacting with antigen, raises the biological problem of self-recognition, since generally all self-molecules can act as antigens when injected into genetically different animals. Although Ehrlich proposed a selective theory of immunity, he did not provide an adequate explanation for the mechanism of self-tolerance. During the 1920s and 1930s, the enormous diversity of antigens and the exquisite specificity of the immune response was recognized, which led to the suggestion that antigen instructed the body to make complementary protein molecules, so as to explain the large number of antibodies which had been identified in various hapten carrier studies (Pauling, 1940). However, these "instructive theories" only provided a mechanism of antibody diversity but did not explain self-tolerance.

The recognition of chimaeric animals (Owen, 1945) led Burnet to suggest that self-tolerance could be acquired during foetal life by inhibition of self-reactive clones (Burnet and Fenner, 1949) and this phenomenon was subsequently confirmed by the extensive experimental work of Medawar and co-workers (Billingham et al., 1956). The enunciation of the central dogma of molecular biology (Crick, 1970) which states that nucleic acid commitment has to precede protein expression has brought back into favour "selective theories" of antibody diversity, which state that the amino acid sequence of the antibody binding site is made before the appearance of antigen. The sole function of antigen in such theories is to select the lymphocytes carrying the immunoglobulin receptors having the best fit and stimulate their proliferation. These two concepts of acquired tolerance and selective theory of antibody diversity, were combined by Burnet into the first truly comprehensive theory of immunity, the elegant and influential "clonal selection theory" (Burnet, 1959). Some parts of the clonal selection theory have been amply confirmed by subsequent experimental studies, such as the selection and proliferation of B-lymphocytes, carrying immunoglobulin receptors of best fit, into antibody-producing plasma cells or the acquisition of a pool of memory cells to explain an altered and enhanced secondary immune response. If a new comprehensive theory of immunity were to emerge, it would still have to incorporate the successful parts of the clonal selection theory.

The scientific method essentially consists of proposing theoretical solutions to problems which can then be tested, first for rational consistency and, second, for empirical content by experimental verification of theoretical predictions. Therefore, the validity of theoretical models can be examined by two clearly separate methods. First, the model solution should be essentially rational and free of logical contradictions when examined for mathematical consistency. Second, the theory or model should be sufficiently explicit to prohibit the existence of certain phenomena which can be checked by appropriate experimental investigations. If such phenomena can subsequently be demonstrated to exist experimentally, then the initial theory has been disproved and a new theory has to be proposed to encompass both the old and new phenomena (Popper, 1962). It is apparent that a theoretical model can be demonstrated to be invalid, before any experimental investigations are undertaken, by the demonstration of an irrational element within the theoretical framework regulating the model.

It is proposed to examine here some of the theoretical implications of the selective theories of immunity by using concepts derived from information theory. If the clonal selection theory of immunity is to be a viable model of the immune response, then it must be consistent with the rules regulating information transfer, and these rules, whether they deal with signals from an emitter to a receiver or from an antigen to an antibody, are derived from information theory.

It would appear that there are serious and fundamental weaknesses within selective theories of immunity, especially with the random somatic mutation model of antibody diversity. It is possible that a search for an alternative theoretical framework may become necessary, to explain both antibody diversity and self-tolerance, with models that not only explain the immunological phenomena but are also compatible with information theory.

2. Selective Theories and Antibody Diversity

The clonal selection theory implies that the formation of immunoglobulin-bearing lymphocytes, precedes the appearance of antigen. Within this selective framework, two types of theories of antibody diversity are generally recognized.

(1) A *random somatic mutation theory*, in which an antibody repertoire is acquired during the lifetime or ontogenetic development of each animal and therefore non-identical or different antibody dictionaries are present in each member of a syngeneic group (Jerne, 1971). Since the production of the antibody repertoire is random, each animal will have its own individual set or dictionary, despite the fact that all these animals may be genetically identical and belong to the same syngeneic population. It is this theory that will be analysed for internal consistency by using information theory.

(2) A *germ line theory*, in which an antibody repertoire is acquired during the phylogeny of the species and identical antibody dictionaries are present in the genome of each cell and therefore of each member of a syngeneic group (Hood and Talmage, 1970).

Both theories of antibody diversity can be examined in terms of their capacity to transmit specific information about the shape and structure of antigens, during an immune response, but the germ line theory does not contain adventitious assumptions beyond those required for the explanation of the general problem of gene regulation and control of protein synthesis.

3. Immune Response as Signal Transmission

A primary immune response implies the specific appearance of substances or cells, which were not present or detectable before the animal was exposed to the specific antigen. Thus a rising antibody titre to some micro-organism is taken to indicate that the vertebrate animal was recently infected or exposed to the antigens of that micro-organism. A rising antibody titre can therefore be considered as a specific signal which has been triggered by the exposure of the vertebrate animal to that antigen and resembles the situation of a radio emitter and receiver, transmitting and receiving a radio signal (Fig. 1). Different antibodies have different

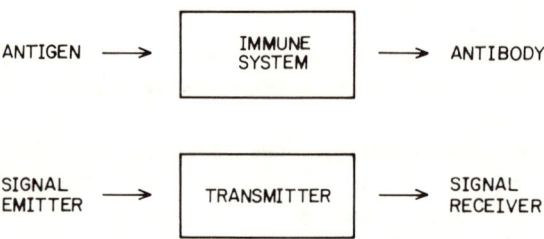

Fig. 1. Antibody can be considered as a signal for the presence, in the not too distant past, of antigen in the body of a vertebrate animal.

three-dimensional complementary cavities, which to some extent define the shapes of antigenic determinants. Since B-lymphocytes are generally considered to produce only one type of immunoglobulin or complementary antibody cavity, only those B-lymphocytes will be stimulated, which have the best fit and, following differentiation into plasma cells, will produce a rising titre against that particular antigen. The B-lymphocytes thus respond in an all-or-none fashion, because they either divide or do not divide, when stimulated by antigen. The presence of specific antibodies in the serum of an animal thus provides information about the stereochemical shape of the antigen, which triggered the immune response. An immune response can therefore be considered as an information transfer system and all such systems can be analysed using information theory (Shannon and Weaver, 1949). When an antibody is detected in serum, there is in that molecule some physical attribute or remnant of the antigen and that is the waveform defining the shape of the antigenic determinant (Fig. 2). Transmission of signals through waveforms is the subject of communications theory and, unambiguously, all such wave transmissions obey the restric-

8. THE RANDOM SOMATIC MUTATION THEORY

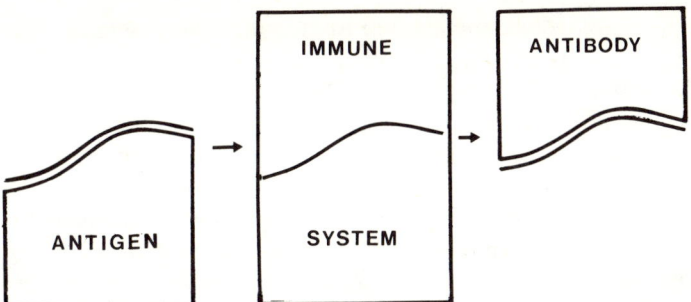

Fig. 2. The three-dimensional shape or waveform of the antigenic determinant is conserved when transferred through an immune system. The study of the transmission of such waveforms is the subject of communications theory.

tions of information theory (Raisbeck, 1963). The transfer of three-dimensional waveforms through an immune system would appear to fall into this category of signal transmission and therefore immunological theories must also be consistent with information theory.

4. Information Theory

Information theory deals with signal sources that possess the ergodic property, in that the code or alphabet used for signal transmission has statistical homogencity over a large number of transmissions (Goldman, 1953). An infinite number of sequences is called an "ensemble" and a sequence of characters, chosen at random, with constant probabilities, constitutes a "stationary source". If probabilities can be assigned such that any statistic obtained by averaging over the "ensemble" does not depend on the distance from the start, then the source is said to be "stationary". When a source is stationary and when every possible "ensemble average" of letters or signals is equal to the corresponding "time average", then the source is ergodic. The "ensemble average" refers to a particular selection of sequences whilst the "time average" is the average of all signals. The theorems of information theory apply to ergodic sources and their proofs rest on the assumption that the message source is ergodic. For an ergodic source, the statistics of a message, for instance the frequency of occurrence of a letter, or a particular signal, does not vary along the length of the message. Furthermore, the probabilities, or description of the source arrived at through the examination of one message, apply equally well to all messages generated by the source and not just to the particular message examined, because the time averages and ensemble averages are the same.

The immunological equivalent of this ergodic requirement is that similar signals or antibodies should have similar ensemble averages such as similar amino acid sequences in the antibody binding sites, when different animals are stimulated by the same antigens. Iso-electrophoretic focusing experiments have shown the presence of similar patterns of antibody molecules, because major and minor clones of immunoglobulins can be readily identified when syngeneic animals are immunized with the same antigens (Schlossman, 1972; Braun *et al.*, 1973). This indicates that different members of a syngeneic population use statistically equivalent codes for the production of an immune response when stimulated by the same antigen and, therefore, the immune process can be considered as an ergodic system.

Information theory is a branch of mathematics which was developed in the first place to explain the amount of information conveyed through radio and telephone signal transmissions. It was observed by Nyquist (1928) that for transmitting telegraph signals at a given rate (signals/second), a definite frequency bandwidth is required.

An early attempt at a definition of information was made by Hartley (1928). From an alphabet of D distinct symbols, D^N different words can be selected, each word containing N symbols. If these words are all equiprobable, then the selection of one word at random is associated with a quantity of information I such that:

$$I = \log D^N$$
$$= N \log D \tag{1}$$

Subsequently the properties regulating information transfer were established on a sound mathematical basis (Shannon and Weaver, 1949; Khinchin, 1957) and generalized for all systems of information. Therefore the principles and theorems of information theory provide testable parameters for the study of information transfer during an immune response.

Information has been defined by Shannon and Weaver (1949, theorem 2) as a probability log function, which has been summed over all the different signals produced by the source:

$$I = -\sum_{i=1}^{n} p_i \log_2 p_i \tag{2}$$

where I is the information of the system and is expressed in binary digits or bits, n is the number of different signals produced by the source, and p_i is the probability of the ith signal or ith letter of the code or alphabet. If natural logarithms are used then information is expressed in natural units or nats.

In the immune system, n would stand for the total number of different amino acid sequences specifying antibody binding sites, encoded in all the lymphocytes and thereby it would define the size of the antibody repertoire in that animal.

This definition of information has a formalistic resemblance to the thermodynamic function of entropy, as défined by Boltzmann's equation (Denbigh, 1966):

$$S = k \ln W \tag{3}$$

where S is the entropy, k is Boltzmann's constant and W describes the number of different states in which the particles of the system may be distributed. However, the precise relation between the thermodynamic function of entropy and the concept of information in information theory is at present under debate (Wilson, 1968).

Information is essentially a probability expression for the uncertainty of the receiver as to what kind of a signal he is going to receive (Brillouin, 1962). Thus messages that are very probable have little information, messages of low probability carry a lot of information and, of course, a totally expected message carries no information at all.

Maximum information will be transmitted when the various probability states are all equally likely (Shannon and Weaver, 1949; Reza, 1961) and then the expression for information acquires the following simple form:

$$I = -\sum_{i=1}^{n} p_i \log_2 p_i = \log_2 n \tag{4}$$

This can readily be seen, when examining a simple source, emitting two separate signals, with probabilities p_1 and p_2, respectively. The information (I) of the source is given by:

$$I = -p_1 \ln p_1 - p_2 \ln p_2 \tag{5}$$

For maximum information

$$dI/dp = 0 \tag{6}$$

and expression (5) becomes

$$I = -p_1 \ln p_1 - (1 - p_1) \ln(1 - p_1) \tag{7}$$

Therefore (6) becomes

$$dI/dp_1 = -\ln p_1 - 1 + \ln(1 - p_1) + 1$$
$$= \ln((1 - p_1)/p_1)$$
$$= \ln(p_2/p_1)$$
$$= 0$$

Therefore

$$p_2/p_1 = 1 \quad \text{or} \quad p_1 = p_2 \qquad (8)$$

For maximum transfer of information from a source having two signals, the probability of each signal is 0.5.

In the general case, for a source of n signals, maximum transfer of information will occur when all signals are equiprobable and the probability of each signal is $1/n$. This simple formula can be used to assess the maximum amount of information that can be transmitted by the immune system, when using different models of antibody diversity.

5. The Random Somatic Mutation Theory of Antibody Diversity

In the random somatic mutation theory, antibody diversity is acquired during the early life of each animal, by random mutations occurring in T- and B-lymphocytes, which have been stimulated to divide either by the animal's own antigens (Jerne, 1971) or by some other hypermutation mechanism (Cunningham, 1976). Antibody diversity is produced by the generation of random mutations in proliferating clones of lymphocytes whose specificity is directed against autologous and homologous transplantation and other self-antigens, and this set of random mutations will vary from one syngeneic animal to another.

Eventually a dictionary of 10^5–10^7 different antibody mutants will be obtained, say, in a mouse, and this number of mutations, it is suggested, will be able to respond to any antigenic determinant the animal is likely to encounter during its lifetime. In the random somatic mutation theory, it is claimed that different syngeneic animals, carrying different random mutations, will still be able to respond to the same antigenic determinants, because of cross-reactivity in the binding sites of mutated lymphocytes. Equation (4) can be used to assess the amount of information that will be transmitted when cross-reactivity is said to occur in different lymphocytes, generated by a random mechanism, in different syngeneic animals, when reacting to the same antigen.

The problem arises as to how specificity can be maintained when antibody diversity is generated by a random mechanism. How can vertebrate animals, belonging to an inbred strain and having identical genes but different random mutations, respond to a specific antigen by producing a similar immune response.

It has been repeatedly observed that when one member of a syngeneic group of animals responds to one specific antigen by an immune response, then every member of that group can also respond to immunization by the same antigen. If an animal does not respond on the first immunization, then it will respond after a second immunization, or a third, thereby indicating that there are specific immunocompetent cells for that particular antigen in every member of that inbred strain. The occasional failure of a primary immune response or the failure of a vaccination "take" is due to technical problems of immunization and not to absence of specific immunocompetent cells. There does not appear to be a specific immunodeficiency that is present in one syngeneic animal and not in a second syngeneic animal, as would be expected if antibody diversity had been produced by random and, therefore, different mutations, in different syngeneic animals.

A random process implies that one animal will make better mutations against one type of antigen compared to another syngeneic animal which will have different mutations, but which in turn may have better mutations or better antibody binding sites against some other antigen. The larger the number of animals studied in a syngeneic population, the greater the diversity of random mutations and therefore the more divergent the antibody responses should be when measured against different antigens. A conservative estimate in the number of possible different mutations is obtained by considering the number of amino acid hypervariable positions in the variable regions of light and heavy chains (Wu and Kabat, 1970). If there are three hypervariable regions in each heavy and light chain and if each hypervariable region consists of at least six amino acids, then the total number of different possible antibody binding sites is of the order of 20^{36}. There are 20 different amino acids occupying 36 different hypervariable positions, 18 in the light chains and a similar number in the heavy chains. The number of possible mutations, say in a 3-month-old mouse, is of the order of 10^8, a rather small number when compared to 20^{36} different possible antibody binding sites.

If a random process is generating antibody diversity, then one BALB/C mouse will produce 10^8 mutations, out of a possible 20^{36} different combinations and a second BALB/C mouse should make 10^8 different random mutations compared to the first mouse and the process of divergence will continue for a third, fourth, fifth mouse and so on. The number of identical mutations or overlap between two BALB/C mice should be extremely small, if we are dealing with random mutations. Since the number of animals in the BALB/C strain or any other syngeneic group can be as large as desired, the diversity of antibody mutations and therefore the diversity of antibody responses, should vary from one mouse to another. This diversity of responses should increase or diverge with the number of animals studied in an inbred syngeneic population. If antibody binding sites vary from one animal to another, we should expect variations in immune responses from one syngeneic animal to another. But this is precisely what does not happen. What is observed is that if one animal can respond to an antigen, then every animal of that strain can respond to that antigen, within well-defined experimental limits. This capacity of every member of a syngeneic group to respond to an antigen, if one animal of such a group is shown to respond to the antigen, this apparent convergence of immune responses, is somewhat of a paradox in the random somatic mutation theory.

The difficult problem to explain is: How can every animal of an inbred group, no matter how large the group, be capable of responding to every antigen, except self-antigens, by a specific immune response, if every animal has acquired a repertoire of random mutations, which differs from the repertoire of every other animal in the syngeneic group? This paradox is not a trivial problem, but a serious weakness in the formulation of the theory.

If one mouse has 10^8 B-lymphocyte mutants out of 20^{36} possible antibody binding sites and a second mouse has another 10^8 mutants, there is no reason whatever to suggest that these random mutations in the two mice are in any way equivalent. There is no *a priori* reason why 10^8 mutations in one mouse should be more or less equivalent to 10^8 mutations in a second mouse, precisely because we are dealing with a random process. It is the randomness itself of the process that ensures that the repertoire in the first mouse will diverge from the repertoire of a second mouse, which will diverge from the repertoire of a third mouse and so on. This in turn will ensure that the immune response of the first mouse to any antigen, must be different from the immune response of every other mouse, coming from the same syngeneic set.

A simple example can illustrate this crucial difference. If a person selects 3 cards out of a pack of 52 cards, then a second person also selects 3 cards out of another full pack of 52 cards, then a third person does the same and so on, one cannot say that the cards selected by each person are more or less equivalent to one another. One person may select three kings and another person may select three jacks or any other combination. There is no convergence in this process of card selection, because the number selected is smaller than the total number of different cards in the pack.

Similarly if one mouse generates 10^8 mutants out of 20^{36} protein possibilities and a second mouse generates another 10^8 random mutants out of 20^{36} possibilities, it cannot be said that all these mutations are more or less equivalent and will enable the animal to respond to any non-self-antigens it may encounter. It is this argument of equivalence or cross-reactivity that represents the fundamental weakness of the random somatic mutation theory (Ebringer, 1975).

6. Information Theory and Cross-reactivity

Let us consider a system of 1000 antigenic determinants and 1000 B-lymphocytes, such that each antigenic determinant can stimulate only one B-lymphocyte. Then according to expression (4), the information (I) generated by this system is given by:

$$I = \log_2 n$$
$$= \log_2 1000$$
$$= 9.97 \text{ binary units (bits)} \tag{9}$$

Let us further consider the possibility that this system cannot generate 1000 B-lymphocytes by random mutation or any other process, but can produce only 100 different B-lymphocytes. If the system is to remain capable of responding to the same number of antigenic determinants as before and do this with a reduced number of lymphocytes, each new B-lymphocyte must become responsive to, or liable to be triggered by, ten different antigenic determinants. If each B-lymphocyte acquires the capacity to respond to ten different antigenic determinants, the system will still be capable of handling a signal from any of the 1000 antigenic determinants as before and the quantity of antibody produced will be the same as on previous occasions. Thus in different syngeneic animals, each generating 100 B-lymphocytes, different random mutations will occur, but as long as one mutant lymphocyte arises out of the ten possible ones, there will always be an antibody produced from this system, when it is stimulated by any one of the original 1000 antigenic determinants.

The system looks as efficient as before, because, irrespective of which one of the 1000 antigenic determinants stimulates the B-lymphocytes, there is always an antibody produced signalling the presence of immune capacity. Superficially the system looks as immunocompetent as before, however, the information transmitted is now given by:

$$I = \log_2 n$$
$$= \log_2 1000/10 = \log_2 100 = 6.64 \text{ bits} \tag{10}$$

where n is the number of different B-lymphocytes, which in this case is 100. So the amount of information obtained per immune response from this system has now been reduced. Although the antigens are bombarding the immune system at the same rate as before, and despite the number of evoked immune responses being the same, the amount of information transmitted across the system has been reduced. The reason why information or specificity has been lost is that when a particular B-lymphocyte has been stimulated, we do not know which one of the ten different cross-reacting antigens has been responsible for this stimulation. The argument can be extended further. If instead of having 100 B-lymphocytes, the system can only generate 10 B-lymphocytes, the information transmitted will be even smaller:

$$I = \log_2 1000/100$$

$$= \log_2 10$$

$$= 3.3 \text{ bit/response} \tag{11}$$

If one B-lymphocyte can be stimulated by 50% of the antigens, the information transmitted decreases further still:

$$I = \log_2 1000/500$$

$$= \log_2 2$$

$$= 1 \text{ bit/response} \tag{12}$$

When any B-lymphocyte can be stimulated by every antigen, that is all antigens are cross-reacting, then the information provided by this system is, as expected, zero:

$$I = \log_2 1000/1000$$

$$= 0 \tag{13}$$

The message is totally expected and therefore conveys no information.

In the model, ten non-overlapping sets of cross-reacting antigen and antibody groups have been considered for purposes of simplicity, but an identical result would obtain if the same degree of cross-reactivity was present in a partially overlapping model. This is actually a closer approximation to the observed behaviour of antibodies in that a single antigenic determinant may stimulate several different B-lymphocytes (Kreth and Williamson, 1973). If antigen 20 evokes antibodies 16–25, antigen 21 evokes antibodies 17–26, antigen 22 triggers antibodies 18–27, and so on, then we have a partially overlapping model, but the average quantity of information transmitted is the same as before, namely $\log_2 1000/10 = 6.64$ bits per immune response, because the degree of cross-reactivity is the same as before.

In the general case, as cross-reactivity increases, there is a decrease in the amount of information or specificity transmitted per immune response (Fig. 3):

$$I = \log_2 n/x \tag{14}$$

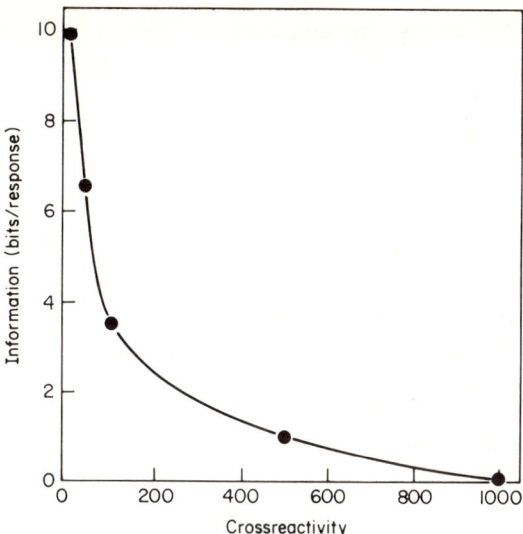

Fig. 3. Information transferred per immune response with varying cross-reactivity in an immune system containing 1000 antigens and an equivalent number of lymphocytes.

where n is the total number of different lymphocytes and x is the average number of different antigens capable of triggering a given B-lymphocyte, the range of x being $1 \leq x \leq n$.

When $x = 1$, that is when there are no cross-reacting antigens, eqn (14) reduces to eqn (4). Only when $x = 1$, is there a one-to-one mapping between antigenic determinants and B-lymphocytes, and in this situation, the maximum amount of information can be transmitted through the system. Since each antigenic determinant stimulates several different lymphocytes, this indicates that the immune system is not operating at the maximum level of information transmission, but at some lower level, which is determined by the total number of major and minor clones evoked by a given antigenic determinant.

If syngeneic animals respond to the same antigens, with the stimulation of presumably different lymphocyte mutations, then, in the random somatic mutation theory, it is suggested that a specific response may still be possible because of cross-reactivity. However, the preceding analysis indicates that specificity cannot be generated by invoking cross-reactivity in the capacity of mutated lymphocytes to respond immunologically, because cross-reactivity leads to loss of information and therefore to loss of immunological specificity. Furthermore, loss of specificity occurs rapidly, even with a small number of cross-reacting antigens, because the rate of informational loss is proportional to the number of cross-reacting antigens.

From (14) and re-expressing in natural logarithms: If

$$I_e = \ln n/x$$

then

$$dI_e/dx = -1/x \qquad (15)$$

where I_e is the information expressed in natural log units or nats, n and x are, as in expression (14), the total number of different lymphocytes and the average number of different antigens capable of triggering a given lymphocyte, respectively. dI_e/dx is the rate of change of information or specificity with respect to the number of cross-reacting antigens. It is interesting to note that the largest falls in dI_e/dx occur for the smaller values of x.

It is clear that as cross-reactivity increases, the immune system becomes less informative or less specific and its capacity to transmit information becomes seriously impaired. It is for this reason that one cannot propose that two BALB/C mice, having two different B-cell repertoires obtained by random mutations, are more or less equivalent to one another, and generate the same specificity when stimulated by the same antigen. Experimentally, however, it is clear that these two mice do produce the same amount of specificity when stimulated by the same antigen. Yet, considered from the random somatic mutation theory, the two repertoires, in these two mice, diverge from one another, and if a third repertoire is added, then the divergence increases further still, and so on. This divergence cannot be compensated by cross-reactivity, when responding to one particular antigen, because specificity will be lost in the process. Since the number of animals in a syngeneic population can be as large as we like, cross-reactivity will be invoked more and more frequently when considering an immune response, with consequent loss of information or specificity. Therefore, as the number of syngeneic animals approaches infinity, cross-reactivity will progressively increase and specificity or mean informative content per animal will approach zero, which is contrary to experimental evidence.

The immune system cannot be built on a model in which specificity progressively falls as the number of animals examined is increased. The essential attribute of the immune system is that it conveys and maintains specificity and therefore it cannot be built on an antibody diversity model, whose inherent property is the destruction of specificity, as the curiosity of the immunologist drives him to examine an increasing number of animals. This *reductio ad absurdum* of a theoretical model invalidates its use as a possible explanation of a biological problem. The presence of a logical fallacy or internal inconsistency within the model proposed for antibody diversity generated by random mutations, means that the theory has failed a scientific test of the first kind (Popper, 1962). Therefore, the random somatic model can no longer be considered as a cogent and viable solution of the problem of antibody diversity.

There is another compelling reason why cross-reactivity cannot go on increasing in an immune system produced by a random somatic diversification process and that is the problem of self-tolerance. If external antigens can stimulate an increasing number of B-lymphocytes through cross-reactivity, then there is no reason why self-antigens should not do the same. Self-antigens could either stimulate the production of autoantibodies or the cross-reactive lymphocytes should have been removed during the presumed phase of foetal elimination of self-reactive clones. Since B-lymphocytes are present in normal vertebrate animals, this is presumptive evidence for a limit to cross-reactivity and, in a sense, this defines the boundaries of specificity in lymphocytes. Such a boundary is reached in a B-lymphocyte when it can no longer be triggered by a particular antigenic determinant. The presence of B-lymphocytes in the absence of autoantibodies is indirect empirical evidence that such cells could not have arisen by a process of random somatic mutation in absence of specific antigen.

7. Linguistics and Cross-reactivity

The paradox of cross-reactivity in antibody diversity can be illustrated by a simple linguistic example (Fig. 4). The word "cour" is the cross-reacting element in this table of French phrases and the set could be considered to represent examples of different mutations in different syngeneic animals.

If one animal produces a mutation equivalent to "la cour de Louis XIV" and a second animal a mutation equivalent to "un cours d'immunologie", these two mutations cannot be used interchangeably in either animal without loss of information, even if they do stimulate the lymphocytes to proliferate and lead to production of antibodies. If the first animal responds to the antigen "la cour de Louis XIV" and also to the antigen "un cours d'immunologie" by producing antibodies of the type "la cour de Louis XIV", then information has been lost, because all that can be said is that the particular word "cour" has been transmitted through this system. The reason why only the word "cour" has been informationally transferred and not the whole antigenic determinant "la cour de Louis XIV", even when complementary antibody "la cour de Louis XIV" has been produced, is because we cannot determine which of the two antigens was responsible

> un cours d'immunologie
> la cour de Louis XIV
> il court d'une femme à une autre
> la basse-cour
> le parcours d'un hippodrome
> un court-de-circuit
> la cour du roi Pétaud
> le cours de la Seine
> la cour de cassation
> le cours de la livre sterling
> le prix Goncourt
> faire la cour à quelqu'un
> le métro Goncourt
> une cour anglaise
> un plat de courgettes

Fig. 4. The word "cour" can be found in many different phrases, each having a distinctive, specific meaning. The cross-reactivity of the word "cour" cannot be used to obtain the specific meaning or informative content of each separate phrase.

for the lymphocyte stimulation and the same argument would apply to the second animal. The greater the number of phrases containing the cross-reacting word "cour", the greater would be the loss in specificity or information.

It is for these reasons that antibody diversity cannot be generated by a random process in absence of antigen, irrespective of whether the mutations are generated by a mutase enzyme mechanism or following stimulation by self-antigens (Jerne, 1971).

8. Summary and Conclusion

The immune response can be considered as a process involving transfer of information because a rising antibody titre can be considered as a specific signal for the presence of antigen, in the recent past, within the body of a vertebrate animal.

Information theory deals with the properties of all systems transferring specific signals and it can be readily demonstrated that the capacity to transmit information decreases rapidly as cross-reactivity increases within a transmitting system.

The random somatic mutation theory is based on the assumption that different random mutations in different animals can still generate a specific immune response because of cross-reactivity in different lymphocytes. This assumption requires a continuous increase in cross-reactivity as the size of the population examined is increased with a consequent loss in the mean capacity to transmit specific information. It is evident that the divergent property of cross-reactivity is incompatible with the convergent requirement for the capacity of the system to maintain a mean specific immune response, as the number of animals studied is increased. A logical fallacy has been demonstrated within the random somatic mutation theory and therefore it is suggested the theory cannot be used as a model for the generation of antibody diversity in vertebrate animals.

It is concluded that the random somatic mutation theory of antibody diversity is untenable, because it is incompatible with information theory.

References

Billingham, R. E., Brent, L. and Medawar, L. B. (1956). *Nature, Lond.* **178**, 514.
Brillouin, L. (1962). "Science and Information Theory." Academic Press, New York.
Braun, D. G., Kjems, E. and Cramer, M. (1973). *J. exp. Med.* **138**, 645.
Burnet, F. M. (1959). "The Clonal Selection Theory of Acquired Immunity." Cambridge University Press.
Burnet, F. M. and Fenner, F. (1949). "The Production of Antibodies." Macmillan, Melbourne.
Crick, F. (1970). *Nature, Lond.* **227**, 561.
Cunningham, A. J. (1976). "The Generation of Antibody Diversity: A New Look." Academic Press, London.
Denbigh, H. (1966). "The Principles of Chemical Equilibrium." Cambridge University Press.
Ebringer, A. (1975). *J. Theor. Biol.* **51**, 293.
Ehrlich, P. and Morgenroth, J. (1901). *Berl. Klin. Wochenschr.* **37**, 234.
Goldman, S. (1953). "Information Theory." Dover, New York.
Hartley, R. V. (1928). *Bell Syst. Tech. J.* **7**, 535.
Hood, L. and Talmage, D. W. (1970). *Science* **168**, 325.
Jerne, N. K. (1971). *Eur. J. Immunol.* **1**, 1.
Khinchin, A. I. (1957). "Mathematical Foundations of Information Theory." Dover, New York.
Kreth, A. W. and Williamson, A. R. (1973). *Eur. J. Immunol.* **3**, 141.
Nyquist, H. (1928). *Trans. Amer. Inst. Elect. Eng.* **47**, 617.
Owen, R. D. (1945). *Science* **102**, 400.
Pauling, L. (1940). *J. Amer. Chem. Soc.* **62**, 2643.
Popper, K. R. (1962). "Conjectures and Refutations: The Growth of Scientific Knowledge." Routledge and Kegan Paul, London.

Raisbeck, G. (1963). "Information Theory." MIT Press, Cambridge.
Reza, F. M. (1961). "An Introduction to Information Theory." McGraw-Hill, New York.
Schlossman, S. F. (1972). *In* "Genetic Control of Immune Responsiveness" (H. O. McDevitt and M. Landy, eds.). Academic Press, London.
Shannon, C. E. and Weaver, W. (1949). "The Mathematical Theory of Communication." University of Illinois Press, Chicago.
Wilson, J. A. (1968). *Nature, Lond.* **219**, 535.
Wu, T. T. and Kabat, E. A. (1970). *J. Exp. Med.* **132**, 211.

Index

Abiotic variables, 194
Age distribution, 2, 19
Age-structure, 90
Alcaligenes faecalis, 100
Anacystis nidulans, 100
Antibody, 287
Antibody diversity, 287
Antigen, 287
Autocorrelation function, 242
Autocovariance function, 241

Backward step operator, 243
Balanced growth, 199
Batch fermenter, 38
Beggiatoa, 208
Bifurcation, 69, 110
Biofilm, 171
Bits, 292
Blasius–Stanton diagram, 227
Boltzmann's equation, 293
Box and Jenkins technique, 240
Branching process, 25
Brownian diffusion, 210
Bulk transport, 180
Butterfly catastrophe, 113

Canonical forms, 112
Catastrophe machine, 106
Cell growth models, 28
Cell quota, 86
Centroid factor method, 141
Characteristic equation, 64
Characteristic response time, 80
Chemical conversion, 181
Chemostat, 41, 77, 91
Clonal selection theory, 288
Coefficient of variation, 5
Coherence number, 81, 99
Colebrook–White reaction, 225
Collins and Richmond equation, 17
Colpidium campylum, 100
Colpoda steinii, 100
Conservation equations, 178

Continuity, 131
Continuous-flow fermenter, 40, 182, 220
Control space, 110, 123
Control variables, 69, 112
Convolution integral, 90
Correlation coefficient, 6
Covariance, 6
Cross-correlation function, 251
Cross-covariance function, 251
Culmination, 130
Cusp catastrophe, 112, 114, 123

Delay, 253
Delay-differential equations, 82
Determinant, 68
Deterministic models, 177
Dictyostelium discoideum, 101, 119, 121
Diffeomorphism, 124
Difference operator, 245
Differentiation, 131
Diffusion, 181, 210
Diffusion barriers, 48
Diffusivity, 186
Digital filtering, 26
Dilution rate, 184
Dimensionless equations, 60
Dirac delta function, 8, 16
Discontinuity, 105
Discrete delay, 87
Divergent oscillations, 81
Doubling time, 39

Eddy diffusion, 210
Eigenvalues, 64, 80, 140
Elliptic umbilic catastrophe, 113
Empirical models, 176
Ensemble, 291
Entropy, 293
Equilibrium, 189
Equilibrium surface, 109
Ergodic property, 291

Expected value, 4
Exponential divergence, 81
Exponential growth, 9, 10, 38
Extensive parameters, 178

Factor analysis, 140
Fick's Law, 49
Flocculation, 100
Fluid drainage force, 210
Focus, 65
Fold catastrophe, 112, 115
Folic acid, 130
Food chain, 91
Fourfold point coefficient, 141
Frequency-domain analysis, 237
Frictional drag forces, 210

Gamma distribution, 184
Gauss–Newton method, 53
Gaussian distribution, 7
Generation rate, 4
Generation time distribution, 4, 6, 8, 12
Germ line theory, 289
Glycocalyx, 171
Grex, 120
Growth models, cell, 28
Growth models, population, 43

Heterogeneous equations, 189
Holding time, 182
Homeostasis, 131
Homogeneous reactions, 189
Hyperbolic umbilic catastrophe, 113
Hyphomicrobium, 208

Immunity, 287
Impulse response model, 239
Information transfer system, 290
Intensive factors, 178

Jacobian, 63
Joint distribution, 5

Klebsiella aerogenes, 101

Laplace transformation, 28
Liapounov's second method, 70
Limit cycle, 67, 79, 81, 82
Linear cell growth, 10

Linear systems, 237
Linearization, 56, 59, 79
Logistic equation, 40
Lotka–Volterra equations, 91, 114, 121, 136
Luxury consumption, 87

Marginal distribution, 6
Marquadts method, 54
Mass balance, 37, 46, 78, 180
Mass transfer, 181
Matrix, 62
Maxwell convention, 127
Mean, 5
Michaelis–Menten equation, 45, 52
Micromixing effects, 186
Mixed process, 244
Moments, 4
Monod kinetics, 42, 84, 92
Monod–Haldane kinetics, 57, 60
Moody diagram, 227
Morphogen, 131
Moving average, 244
Multiple substrate kinetics, 43

Natural units, 292
Nitrification, 193
Node, 65
Noise, 238, 243, 252, 274
Normal distribution, 7

Operating diagrams, 97
Overdamped systems, 80

Parabolic umbilic catastrophe, 113
Parameters, 42
Partial autocorrelation function, 244
Pearson product moment correlation, 142, 165
Pearson type III distribution, 29
Perfect delay convention, 126
Periodic behaviour, 248
Perturbation variables, 57
Phase plane, 66
Phase space, 109
Phytoplankton, 47
Plug flow reactor, 185
Poincaré–Bendixson theory, 68
Poisson distribution, 28
Population balance models, 176

INDEX

Population growth models, 43
Portmanteau test, 270
Predator-prey interactions, 72, 82, 121
Prewhitening transformation, 274
Primary immune response, 290
Primary wave, 130
Principal factor method, 140
Probability distribution, 3
Process analysis, 174
Process rates, 180

Random somatic mutation theory, 289
Rate equation, 187
Relaxation time, 39
Renewal equation, 25
Residence time, 182, 184
Response time, 80

Scientific method, 288
Secondary wave, 130
Sedimentation tank, 261
Segregated models, 177
Segregation, 186, 199
Self-tolerance, 287
Sewage treatment, 261
Simplifying assumptions, 197
Single-series analysis, 240, 241
Single-series models, 263
Singularity, 105
Size distribution, 2
Skewness, 5
Sphaerotilus, 208
Splines, 259
Spore cells, 120
Stability, 67
Stability analysis, 79
Stability matrix, 80
Stalk cells, 120
State variables, 42, 69, 112
Stationarity, 241
Stationary source, 291
Stationary time series, 241

Steady-state, 8, 56, 77, 79, 189
Steepest descent method, 53
Stiff equations, 51
Stochastic models, 177
Stoichiometry, 46, 190
Structural stability, 113, 131
Structured models, 44, 176, 197
Substrate inhibition kinetics, 70, 115
Surface roughness, 211
Swallowtail catastrophe, 113
Synchronous cultures, 22

Taylor series, 56, 114
Tetrahymena pyriformis, 101
Thermodynamic properties, 178
Thermophoresis, 210
Time-domain analysis, 237
Time series, 235, 241
Trace, 68
Transfer-function analysis, 240
Transfer-function models, 251, 274
Transformation processes, 181
Transport coefficients, 178
Transport phenomena models, 176
Transport terms, 47
Trickling filter, 171
Turbulent flow, 208

Underdamped systems, 81
Universal unfolding, 114

Variance, 140
Variance reduction, 240
Vectors, 62
Vortex point, 65

Wall growth, 100
White noise, 243, 252

Yule–Walker equations, 245

Zeeman catastrophe machine, 106
Zooglea ramigera, 204